C0-AKT-551

Spectroscopy

Spectroscopy

D. H. Whiffen

Second Edition

Longman

LONGMAN GROUP LIMITED
London

Associated companies, branches and representatives throughout the world

First published 1966
2nd Edition, in SI units, 1971

ISBN 0 582 44656 2

Printed in Great Britain by
The Camelot Press Ltd., London and Southampton

Contents

Foreword

Spectroscopy and its applications form a significant part of modern chemistry and physics. It is comprised of a number of specialist fields which can, to some extent, be treated individually, but for the beginner the fundamental concepts are more readily understood when they are seen in the context of the whole topic. This belief has prompted the author to write this book.

This is therefore the author's attempt to encompass in less than 200 pages all the essentials to understanding spectroscopy and its subdivisions at an undergraduate level. Two initial chapters are concerned with energy levels, radiation and their interconnection. These are followed by chapters on the separate techniques in order of increasing frequency. This the author believes is the best sequence, even though it reverses the path of history. Thus one can start with the simple two-level system of a hydrogen nucleus in a magnetic field.

The book aims to cover all the important principles, the main features of experimental technique, some examples of applications to chemistry and sufficient introduction to the nomenclature to make more advanced texts comprehensible. Much has been intentionally omitted including almost all proofs or derivations of quantum mechanical formulae. Nor has comprehensiveness seemed important.

Any student is advised to read the book through fairly quickly so as to get the whole subject in perspective and to leave until a slower second reading those sections he finds troublesome. During this second reading he may try the problems, which are not too easy and which were intended to be answered from this book alone. But if difficulties should lead the student to consult the more specialist works mentioned in the bibliography, these will introduce him to many other fascinating things which form the science of spectroscopy.

If I do not thank by name any of my scientific friends, a term which embraces teachers, colleagues and students, it is because they are so numerous. I am nevertheless very aware of the great debt I owe them for the part they have all played, and are still playing, in my own education in spectroscopy.

February 1966 DAVID H. WHIFFEN

Note added March 1970

Since the first printing in 1966, there have been strong moves towards the universal acceptance of SI units and also of recommended symbols. The practice recommended in *Symbols, signs and abbreviations recommended for British scientific publications*, The Royal Society, London 1969, has been followed for this reprinting. The text is otherwise essentially unaltered.

In many changes, such as MHz for Mc/s, no serious unfamiliarity arises although some readers may be less happy with the change to joules from calories. The author's own greatest reluctance was associated with the use of p rather than μ for an electric dipole moment. Consistency of units requires the magnetic resonance equations to be written in terms of the magnetic induction or flux density, B, rather than the field, H. Algebraic expressions have been modified to rationalized mksA forms by appropriate insertion of multiples of π and of μ_0 and ε_0, the permeability and the permittivity of a vacuum.

The author apologizes for any errors introduced in making these changes and would be glad to be informed of any detected by readers.

D.H.W.

Energy levels 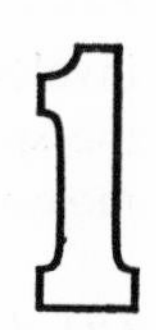

Introduction

From its derivation the word spectroscopy appears to mean the watching of images, but the modern subject covers the interaction of electromagnetic radiation with matter. The most important consequence of such interaction is that energy is absorbed or emitted by the matter in discrete amounts or quanta. A measurement of the radiation frequency gives a value for the change of energy involved and from a complete investigation it is possible to infer the set of possible discrete energy levels of the matter which is being studied. The ways in which the measurements are made and the energy levels deduced constitute the practice of spectroscopy. Use of information about the energy levels constitutes the major applications. The background theory, experimental aspects, interpretational methods and a range of applications are covered in this book.

From the first paragraph it is apparent that spectroscopy is closely linked with quantum theory; some knowledge of this subject is presumed in the reader. This does not mean that familiarity with the mathematical techniques is expected. In many sections formulae based on quantum theory are quoted without proof and reference should be made to standard texts for the derivations. This first chapter is concerned with energy levels. Many topics mentioned in it will be treated in more detail in later chapters.

The more fundamental parts of spectroscopy deal with isolated systems and these may be divided into three essential kinds:

(i) Individual nuclei and fundamental particles. The internal energy levels of these species are very widely separated and the techniques used to investigate them do not use electromagnetic radiation predominantly. Although important experiments are carried out involving γ-ray emissions the results of these experiments cannot be isolated from many others involving interactions between heavy particles. This book does not treat the spectroscopy of nuclear transformations, except in relation to the molecular aspects of the Mössbauer effect.

(ii) Atoms Free atoms, and likewise monatomic ions, possess a set of electronic energy levels which can be determined. Besides differing in energy these states differ in the distribution of electrons around the heavy, positively charged nucleus, with a consequent difference of electrostatic interaction between the charged particles and of kinetic energy of the electrons. Atomic spectroscopy played a very important part in the history of the quantum theory. But the number of atoms known is limited and ions with more than three or four charges are difficult to obtain and the majority of important atomic spectra have already been measured, fully tabulated and interpreted. Interest is chiefly confined to the practitioner and the quantum theoretician, although the subject has considerable importance in astronomy and in analytical chemistry. Applications in the field of chemical physics are few and interest in atomic spectroscopy had declined relatively in recent years until revived by the development of lasers. Most of the general principles apply equally to molecular and atomic spectroscopy and should be taken to do so if the context will allow, even though the relevant section of this book is written in terms of molecules.

(iii) Molecules By contrast to atoms, the number of known molecules is extremely large and each possesses a diversity of energy levels. Current interest in molecular spectroscopy is very great, and an attempt has been made in this book to illustrate the subject in all its aspects.

Molecular energy states

An isolated molecule is said to be, temporarily, in a particular energy state if it possesses the amount of energy used to specify the state concerned.

If a molecule is isolated for a time Δt, the Heisenberg uncertainty principle allows its energy, E, to be measured with a precision $\Delta E \sim \hbar/\Delta t$. Here $\hbar$ is Planck's constant[1] h divided by 2π and is numerically 10^{-34} m^2 kg s^{-1}. If $\Delta t > 10$ s the precision in energy is as good as can be

[1] Recent literature is divided between the use of $\hbar$ and h in standard formulae and derivations. For the most part the use of $\hbar$ gives formulae which are simpler to remember and would always be preferable if frequencies were given in radians/sec. This is not common practice and consequently h and $\hbar$ are both used in this book according to convenience. Although somewhat illogical this procedure does give a student familiarity with both schemes.

measured with the most favourable experimental conditions. Indeed for many purposes isolation for 10^{-10} s is sufficient to allow an accurate and meaningful description of the energy state. Isolation for these times is readily obtainable in low-pressure gases and equivalent effective isolation for the purpose in hand is found in condensed phases, and it is usually legitimate to consider energies as precise quantities which may be accurately measured and used to label energy states. Considerations involving the uncertainty principle are important in discussions of spectral line widths.

To allocate numerical values to the energy it is necessary first to define an energy zero. This choice is in essence arbitrary and is normally made on the basis of convenience. For atomic and molecular problems it is usual to define the state of zero energy to be the *ground* state. This is the state which is occupied to a progressively larger extent as the material in thermal equilibrium is cooled towards the absolute zero of temperature. The nuclei and electrons which form a molecule are only capable of reaching configurations of lower energy than the ground state as the result of chemical reactions. If chemical reactions, including dissociation, are under discussion a self-consistent energy zero must be used; this would normally be the ground states of *either* the reactants *or* the products.

The ground state must be contrasted with the *excited* states which are necessarily of higher energy. A spectroscopic transition may take a molecule between its ground state and an excited state or between two excited states, one of which will be of greater energy and called the upper state in contrast to the lower energy state of the transition. It is customary in describing a transition to write the upper state, say A, first and the lower state, B, second and to include an arrow to indicate the direction of transition. Thus $A \rightarrow B$ implies emission of radiation, the molecule passing from state A to B, as opposed to absorption, which would be abbreviated to $A \leftarrow B$. $A \leftrightarrow B$ implies the transition in either direction. If it is desired to distinguish any property of the upper state from the corresponding property of the lower state the symbol for the property is given a single prime in contrast to a double prime for the lower state. Thus the energies might be written E_A and E_B or more often E' and E'' respectively.

It frequently happens that two or more energy states of a molecule have numerically the same value of the energy, when they are said to be *degenerate*. The states are distinguishable in the sense that some very weak external influence, for example a magnetic field, can be found which acts differently on the states and leads to different modified energies. Such an influence is said to lift, raise, split, remove or resolve the degeneracy; the different verbs are used interchangeably. For many purposes degenerate states must be counted individually. In specific cases it is possible to affirm which types of influence are capable of resolving the degeneracy and which are inherently incapable of doing so. Some are able partially to resolve a degeneracy, for example to resolve a four-fold degeneracy of states into two pairs of doubly degenerate states. The term energy level may be used to refer to a set of degenerate energy states, although this distinction is not universally applied and state and level are often used interchangeably.

Degeneracies are normally inherent in the style of a problem and their existence is not dependent on the precise molecule under discussion. However, an equality of energies may arise because of some precise numerical relationship between some molecular parameters. Such occasional, relatively unpredictable degeneracies are said to be accidental. For instance a degeneracy normally exists because the energy of a rotating molecule is independent of the axis of rotation in space. If additional degeneracy should arise because of an atypical relationship between the moments of inertia, it would be accidental.

Although energy states are properties of individual systems, it is often convenient to consider how they would change if some molecular parameter, such as an internuclear distance, were to be varied continuously. Any small parameter variation will slightly vary each energy state. A large variation can be built up from a large number of small variations and the change of any one energy state followed throughout. There is always a one to one correspondence of states between the initial and final arrangement, which may be as different as a diatomic molecule and its constituent atoms.

It will often happen that an accidental degeneracy occurs at some value of the varying parameter. If minor terms are omitted from the calculation, the curves of energy against parameter will cross as shown by the dotted lines of Figure 1.1, but if more exact considerations are applied some small influence is introduced which removes the degeneracy so that the full curves are followed. This illustrates the non-crossing rule for energy states. In more sophisticated language, two energy states may only cross if they have different symmetries and if perturbations which can mix these symmetries are absent so that the accidental degeneracy at the crossing point is not removed. Care must be taken in correlating energy states, since the state at *a* correlates with *b* if the full curves are followed but with *d* if the dotted portion is taken. If the parameter should be a property which can in practice be varied smoothly by the molecule itself (e.g. an internuclear distance in contrast to an atomic mass) it will follow path *a* to *b* if its motion through the crossing point is slow, but

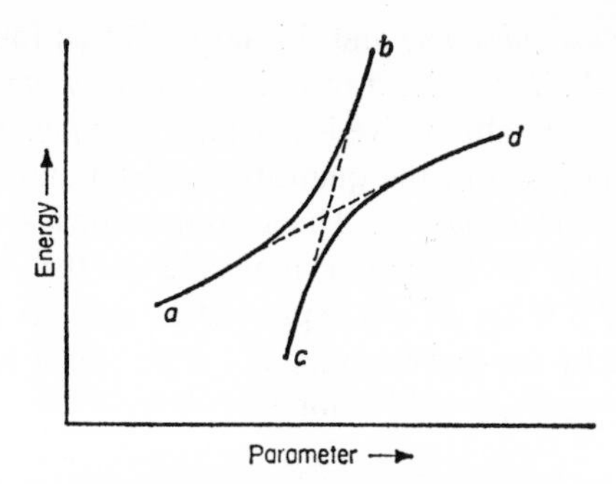

1.1 Non-crossing energy curves.

will follow *a* to *d* if the crossing is reached rapidly so that the kinetic energy greatly exceeds the energy gap between the full curves.

Thermal equilibrium

Spectroscopy concerns the passage of molecules from one energy level to another and this is not an equilibrium process. Nevertheless the disturbance of thermal equilibrium by the spectroscopic investigation may be slight and it is important to know the distribution of molecules, in equilibrium, amongst their possible energy states. The number of molecules in an energy state directly affects the observed intensity of spectroscopic transitions.

The Boltzmann law of energy distribution in its quantum mechanical form requires that, in thermal equilibrium with a thermostat at a temperature T, the relative probability of a molecule being in the ith energy state with energy E_i, above the ground state, is $\exp(-E_i/kT)$ where $k = 1.4 \times 10^{-23}$ J K^{-1} is the Boltzmann constant. It is convenient to normalize the probability so that the molecule is certainly present, which means that the absolute probabilities summed over all energy states is unity. This is true if the absolute probabilities are

$$\exp(-E_i/kT)/\Sigma_j \exp(-E_j/kT).$$

A similar expression can be written in terms of energy levels, namely the absolute probability of finding a molecule in a level of energy E_j and degeneracy g_j is

$$g_j \exp(-E_j/kT)/\Sigma_i g_i \exp(-E_i/kT).$$

Remembering the distinction between states and levels given above, it can be seen that the denominators are equal. These summations are important quantities and are given a special name and symbol. In German the regular name is *Zustandsumme* and this term is often found untranslated in works in English. The strict translation, *sum over states*, is sometimes used, while a common alternative is *partition function*. This suffers from being a poor description of the quantity involved and liable to be confused with partition coefficient. The recommended symbol is Q. The absolute probability of finding a molecule in the ith state is then $Q^{-1} \exp(-E_i/kT)$. If the E_i are expressed as energies per mole then k must be replaced by the gas constant $R = N_A k = 8{\cdot}3$ J K^{-1} mol^{-1}. N_A is the Avogadro constant, 6×10^{23} mol^{-1}.

Relaxation

Occasionally thermal equilibrium is disturbed and is only re-attained in a time comparable with the time of experimental measurements. It is then necessary to have a knowledge of the relaxation time which describes the rate of re-establishment of equilibrium.

Consider the simple case of only two energy states i and j with initial non-equilibrium molecular populations n_i^0 and n_j^0. From the Boltzmann distribution the final population of state i, if the molecules are in contact with a large thermostat at temperature T, is given by

$$n_i^\infty = (n_i^0 + n_j^0) \exp(-E_i/kT) [\exp(-E_i/kT) + \exp(-E_j/kT)]^{-1}$$

and similarly for n_i^∞.

If k_{ij} is the rate constant for transitions from $i \to j$ and k_{ji} is the constant for the reverse direction, the rate of change of n_i with time is

$$\mathrm{d}n_i/\mathrm{d}t = -k_{ij}n_i + k_{ji}n_j.$$

At thermal equilibrium this net rate must be zero and so

$$k_{ij}n_i^\infty = k_{ji}n_j^\infty$$

and thence

$$k_{ij}\exp(E_j/kT) = k_{ji}\exp(E_i/kT) = \tau^{-1}[\exp(-E_i/kT) + \exp(-E_j/kT)]^{-1}$$

where the last expression serves to define τ. Remembering that

$$n_j = n_i^0 + n_j^0 - n_i$$

since the total number of molecules is fixed, the rate equation may be rewritten

$$\mathrm{d}n_i/\mathrm{d}t = \tau^{-1}(n_i^\infty - n_i),$$

whose solution is

$$(n_i - n_i^{\infty}) = (n_i^0 - n_i^{\infty}) \exp(-t/\tau).$$

This expression shows that n_i changes exponentially towards its equilibrium population and that τ is the time constant. If i and j are exchanged τ is unaltered and this shows that the equilibrium is approached from either direction with the same time constant. τ is called the thermal relaxation time.

If more than two states are involved, so that equilibrium is established partly via paths such as $i \rightarrow k \rightarrow j$, the analysis is more complicated and the overall law not necessarily exponential. However, in many practical cases equilibrium is reached exponentially within experimental error and a single relaxation time is an effective parameter to describe the process. The value of τ may be anything from about 10^{-12} s to several minutes according to the nature of the states involved.

Energy Scales

In different branches of spectroscopy different customs have grown up for the use of particular units for expressing quantitative values of molecular energies. The most fundamental unit is probably J/molecule, but this is awkward in practice as large negative powers of ten are regularly required. J mol^{-1} are preferable in this respect, especially since the larger molecular energies are then conveniently described in kJ mol^{-1}. An even larger unit is the electron volt (eV), that is the energy acquired by an electronic charge falling through a potential difference of one volt. It is useful to memorize the equivalence factor, 1 eV = 100 kJ mol^{-1}.

A frequency multiplied by Planck's constant has the dimensions of energy. It can be very convenient to refer to frequencies as if they were energies, it being implicitly understood that the frequency multiplied by h is the energy which is intended. For small molecular energies the units Hz, i.e. cycles/sec, kHz = 10^3 Hz, MHz = 10^6 Hz and GHz = 10^9 Hz, make a convenient set whose interrelation is easily remembered. For the higher molecular energies these units are inconveniently small and the wave number, which is the frequency divided by c the velocity of electromagnetic radiation, is preferred. The common unit is written cm^{-1} and is read aloud as centimetre to the minus one or reciprocal centimetre. The symbol σ is recommended for a parameter denoting wave numbers. Thus $\sigma = c^{-1}\nu$. It is also the reciprocal of the wavelength in vacuo. 1 $\text{cm}^{-1} \equiv 3 \times 10^{10}$ Hz. Further, 1 $\text{cm}^{-1} \equiv 12$ J mol^{-1} and this conversion factor is also worth remembering. More exact values of fundamental and other constants are tabulated on p. 199.

Classification of energies

The total number of energy levels of any molecule is extremely large and some simplifications are essential. For most purposes it is permissible to treat a molecule as if it possessed several distinct reservoirs of energy. The total energy may then be apportioned between the different reservoirs according to an equation such as

$$E_{\text{total}} = E_{\text{translation}} + E_{\text{nuclear orientation}} + E_{\text{rotation}} + E_{\text{vibration}} + E_{\text{electronic}}.$$

Each E in this equation represents the energy pertaining to motions and forces of the types indicated by its subscript.

Translational energies There is a strict solution of the quantum mechanical equations of motion for an isolated molecule of mass m confined to a rectangular box of dimensions $a \times b \times c$, namely

$$E_{\text{translation}} = (h^2/8m)[(n_x/a)^2 + (n_y/b)^2 + (n_z/c)^2].$$

In this the n are restricted to integer values. However, for containers of practical dimensions the number of levels is exceedingly numerous and the intervals between levels negligible, so that an essentially continuous range of translational energies are available. Also intermolecular collisions are normally of more importance than wall collisions, so that the derivation of the above equation is inappropriate. For most purposes translational energy is best treated in a non-quantal or classical manner, that is

$$E_{\text{translation}} = mu^2/2,$$

where u is the velocity of the centre of mass and all values of u are allowed.

Translation corresponds to three degrees of freedom so that by the classical theorem of the equipartition of energy the average value of $E_{\text{translation}} = 3kT/2$ which is about 4 kJ mol^{-1} at room temperature. In solids the translational freedom is restricted and the equivalent motion must be treated in a manner appropriate to co-operative vibrations.

Nuclear orientational energy For molecules containing nuclei which possess nuclear spin the energy may depend on the orientation of the nucleus. The energies involved are always extremely small. The states may be characterized by a quantum number M_I for each such nucleus. M_I represents the projection of the nuclear spin angular momentum, I, in units of $\hbar$ along some specified direction. I is restricted to be an integer or half-integer according to the nucleus in question and M_I may have each of the values $+I, +I-1, +I-2, \ldots, -I+1, -I$ so that the energy states are $(2I+1)$ in number. The energy values depend on the environment of the nucleus and the magnetic fields present and will be discussed more fully in chapters 3 and 4. At all temperatures more than a degree or so above the absolute zero $E_{\text{nuclear orientation}} \ll kT$, $\exp(-E/kT) \simeq 1$ and all orientational states are equally populated to a high degree of accuracy.

Rotational energy In the gas state molecular rotational energies are also quantized and for the lighter molecules the energy separations are $\sim 10^{-1}$ J mol^{-1}. The quantum mechanical formula for the energy of a simple linear molecule of moment of inertia I is

$$E_{\text{rotational}} = J(J+1)(\hbar^2/2I).$$

In this formula J is zero or a positive integer called the rotational quantum number. Each such energy level has a $(2J+1)$-fold degeneracy; the individual states with the same value of J may be characterized by different directions of the rotational axis in space. States up to moderate values of J are appreciably populated at room temperature.

This type of formula does not apply in the liquid phase where molecular collisions are frequent. Instead the rotation is better treated in a classical manner analogous to translations. In most, but not all, crystalline solids free rotation is prevented and replaced by an oscillatory motion which can be treated as a vibration, although there are special features arising from the co-operative motions of adjacent molecules.

Vibrational energy Vibrational energy is comparatively independent of state of aggregation and must be treated on a quantum mechanical basis. Each degree of freedom is treated as a separate harmonic oscillator for which the permitted vibrational energies are given by

$$E_{\text{vibrational}} = h\nu(v+1/2).$$

In this ν is the vibrational frequency and v is zero or a positive integer called the vibrational quantum number. There is no inherent degeneracy unless the molecules are of such a symmetrical shape that two or three degrees of freedom have identical values of ν. It is to be noticed that in this form the formula does not fulfil the convention that the energy shall be zero for the ground state. The ground state corresponds to $v=0$ with an energy $h\nu/2$; this is often called the zero point energy. The origin corresponds to a molecule with all its nuclei at their equilibrium positions and devoid of relative motion.

In any computation the zero point energy must be consistently included or excluded as is most convenient.

The $h\nu$ vary from about 1 to 40 kJ mol^{-1} according to the relevant degree of freedom. At room temperature the $v=1$ and $v=2$ states of low-frequency vibrations are appreciably populated, but for the higher frequency vibrational degrees of freedom practically all molecules have $v=0$ at equilibrium.

Electronic energy Molecules, like atoms, have discrete, quantized excited electronic states which differ in the distribution of their electrons. The energy intervals are many kJ mol^{-1} so that all molecules are in their ground electronic states at equilibrium at room temperature. Different electronic states may differ appreciably in their associated moments of inertia and vibrational frequencies so that the permitted values of $E_{\text{rotational}}$ and $E_{\text{vibrational}}$ are considerably altered by electronic

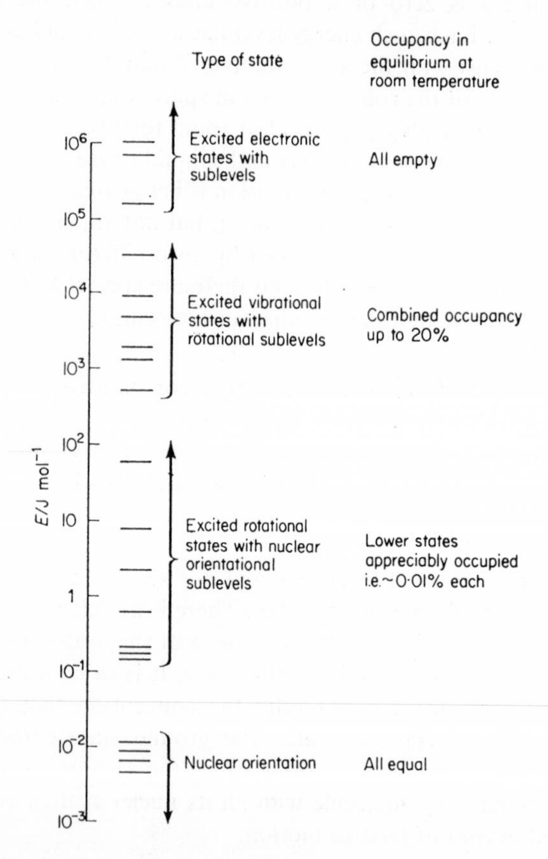

1.2 Summary chart indicating types of energy levels and the extent to which they are filled in equilibrium at room temperature.

excitation. There is no general formula for electronic energies in terms of simple quantum numbers, except for hydrogen-like atoms.

Other energy In special cases other energy is involved which is not easily apportioned under the above headings. The more important energy types not referred to already are associated with the orientation of electron spins and orbital motions, hindered internal rotation, ring puckering motions in some cyclic molecules, inversion, and ionization.

Figure 1.2 summarizes many of the features discussed above, although translational energies are not shown. Few spectroscopic processes involve change of translational energy and for many processes the existence of such energy may be ignored. It must be strongly emphasized that the subdivision into different types of energy is made for purposes of convenience of discussion and ease of understanding. It is only approximately valid and for work of a precise nature adjustments must be made to allow for the defects of simpler treatments in this respect.

2

Electromagnetic radiation and spectroscopic transitions

Wave nature of radiation

Since the changes of energy levels which are studied by spectroscopic techniques are concerned with electromagnetic radiation, it is necessary to mention next the more important properties of such radiation. Electromagnetic waves possess a frequency, ν, and a wavelength, λ, whose product is the radiation velocity. In a vacuum this is always $c \simeq 3 \times 10^8$ m s^{-1}, the velocity of light. In homogeneous media the phase velocity is reduced by the refractive index, n. In general then

$$n\nu\lambda = c.$$

In a medium of constant refractive index it is the wavelength which is reduced by the factor n, whereas the frequency is unchanged. The frequency is then the most important characteristic of the radiation and is the property used to describe it. In the high-frequency spectral regions it is more convenient to measure the wavelength and to use this as the specification: the refractive index of air is nearly one and except for the most accurate work air and vacuum wavelengths are used interchangeably. Visible light is the best known form of electromagnetic radiation, although it is not of special importance in spectroscopy.

Under some experimental conditions only one frequency is present in a system and the radiation is said to be *monochromatic*. In free space, that is in the absence of reflecting walls at distances as small as the wavelength, radiation possesses a definite direction of propagation. As the name electromagnetic indicates there is associated with the radiation—or perhaps one should rather say the radiation consists of—an electrical field, E, and a magnetic field, H. For radiation travelling in the z direction in free space these vary with time and distance as

$$E_x \cos 2\pi(\nu t - z/\lambda)$$

and

$$H_y \cos 2\pi(vt - z/\lambda).$$

The x and y subscripts indicate that the electric and magnetic fields are mutually perpendicular to each other and to the direction of propagation, z. If the whole beam of radiation is of just this form it is said to be *coherent* and *plane polarized.* Two beams, or two parts of the same beam, are said to be incoherent if the phase of one is unrelated to that of the other. No interference effects are obtained between incoherent beams which normally have separate origins. The plane polarized nature refers to a special feature of the radiation, namely $E_y = H_x = 0$ in the above example. The electric vector is confined to the xz plane and this is then the plane of polarization, which necessarily includes the direction of propagation. The sum of two polarized beams of the same phase and the same direction of propagation is also a polarized beam with an intermediate plane of polarization and likewise any beam polarized in a general direction can be expressed as the sum of two beams polarized in specified directions.

Unpolarized light has constituents of its electric field pointing randomly in directions perpendicular to z. However, if there is a phase coherence between E_x and E_y a further special case arises. In particular if

$$E_x = E_0 \cos 2\pi(vt - z/\lambda)$$
$$E_y = -E_0 \sin 2\pi(vt - z/\lambda),$$

the radiation is said to be right *circularly polarized.* There is a complementary situation in which the electric vector appears to turn anticlockwise and the radiation is left circularly polarized. It would be represented by

$$E_x = E_0 \cos 2\pi(vt - z/\lambda)$$
$$E_y = E_0 \sin 2\pi(vt - z/\lambda).$$

It can be seen that the sum of coherent right and left circularly polarized radiation forms plane polarized radiation. Intermediate cases between these extremes are called elliptically polarized radiation and partially polarized radiation. Visible light is polarized by means of a Nicol prism or a polaroid sheet. Circularly polarized light is obtained when plane polarized light is passed through a quarter wave plate set at the correct orientation.

Bohr rule and particle nature of radiation

Quantum mechanics suggests that electromagnetic radiation has a dual nature and when it interacts with matter its particle-like properties appear more prominently. In the early days of the quantum theory it was established that radiation could behave as though it exists as discrete packets of energy which are called *photons* or *radiation quanta.* These have no rest mass and an energy related to the frequency by Einstein's equation

$$E = h\nu.$$

When molecules absorb or emit radiation they absorb or emit exactly one quantum of energy. Consequently E must equal the difference of two energy levels of the molecule, that is

$$E' - E'' = h\nu.$$

This is the *Bohr frequency rule* and is the basis of all quantitative spectroscopy. To the extent that E' and E'' are essentially precise quantities ν will be monochromatic. Radiation of a frequency which is not related to any interval between molecular energy levels in this way will not be absorbed or emitted except under the unusual conditions required for a two quantum transition. Raman scattering is another process for which this statement needs to be modified. The Bohr rule also shows that if the energies are expressed in frequency units as explained in chapter 1, then the difference in the molecular energy levels is numerically equal to the frequency of the radiation: the convenience of using frequency as an energy scale derives from this equivalence. The converse of the Bohr rule is not a requirement of nature and even though $h\nu$ corresponds to an energy interval in the molecule, such radiation may not be absorbed or emitted.

A full discussion of the last point involves transition probabilities and selection rules (see below). There is also for any individual molecule the requirement that it be in the requisite initial energy state. It may have reached this level by the normal mechanisms for the maintenance of the Boltzmann population distribution in thermal equilibrium. Alternatively the initial energy state may be reached as the result of a non-equilibrium process such as an electric discharge.

In so far as the total molecular energy can be split into different types of energy, the spectroscopic transitions can involve changes of energy predominantly of one type. For different types different ranges of radiation frequency are required and the detailed techniques are very different. Figure 2.1 summarizes these ranges showing the corresponding wavelengths and the common name of each of the different frequency regions. The boundaries are not sharply defined. It can be seen that visible light occupies only a very small fraction of the range of interest.

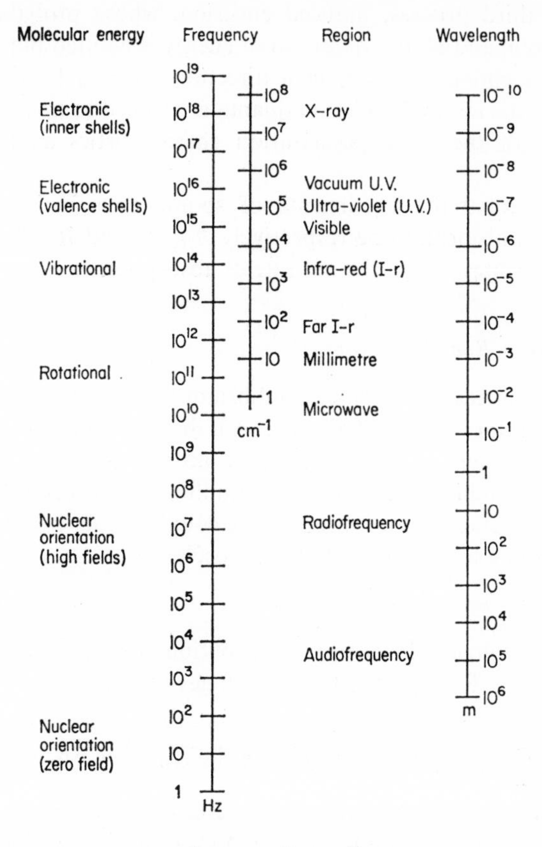

2.1 Nomenclature of various frequency regions and the types of molecular transition with which they are associated.

Spontaneous emissions, absorption and induced emission

Two essential processes can occur in a spectroscopic transition. Either existing radiation can be absorbed with the consequent elevation of the molecule to a higher energy level or else a molecule in an excited state can lose energy and a quantum of radiation is created and this carries away the energy. The probability of the absorption process is directly proportional to the radiation energy density of the requisite frequency, $\rho(\nu)$, and to the number of molecules, n_i, in the lower state. The spontaneous emission probability is proportional to the number of molecules, n_j, in the upper state. Both classical analogy and quantum theory demand a third process, induced emission, whose probability is proportional to n_j and $\rho(\nu)$; a quantum of energy is emitted but whereas the spontaneous emission occurs in a direction dictated by the molecular orientation, the induced emission quanta are forced to travel in the same direction with the same polarization characteristics as the inducing radiation.

Let the probability coefficients of spontaneous emission, induced emission and absorption be respectively A_{ji}, B_{ji} and B_{ij}. For transition between the states j and i the net rate of absorption of radiation is given by

$$B_{ij}n_i\rho(\nu) - B_{ji}n_j\rho(\nu) - A_{ji}n_j.$$

For the special case of thermal equilibrium not only between molecules and a thermostat at T K but also between these and a source of radiation also at T K, there can on balance be no net absorption, since such absorption would imply a thermal disequilibrium between the source and the molecules, such that a source at T K could transfer heat to a thermostat at the same temperature. For this special case

$$B_{ij}n_i\rho(\nu) - B_{ji}n_j\rho(\nu) - A_{ji}n_j = 0.$$

Using the Boltzmann energy distribution for the molecules, i.e. $n_i = Q^{-1} \exp(-E_i/kT)$, etc., this rearranges to give

$$\rho(\nu) = \frac{A_{ji}}{B_{ij}\exp(E_j - E_i)/kT - B_{ji}} = \frac{A_{ji}}{B_{ij}\exp(h\nu/kT) - B_{ji}}.$$

Now the variation of $\rho(\nu)$ with temperature is well known for one source, namely the black body, and this is embodied in Planck's law

$$\rho(\nu) = (8\pi h\nu^3/c^3)[\exp(h\nu/kT) - 1]^{-1}.$$

These two equations for $\rho(\nu)$ as a function of ν and T can only be reconciled provided

$$B_{ij} = B_{ji} = (c^3/8\pi h\nu^3)A_{ji}.$$

This important result shows that if one of the probability coefficients is known the other two are also known. The result can be derived more rigorously from the quantum theory of radiation. The condition $B_{ji} = B_{ij}$ also follows from the principle of microscopic reversibility.

An interesting practical scheme is to measure B_{ij} in the laboratory and use it to compute A_{ji} and thence find the number of emitting centres in astronomical sources. It can also be seen that A_{ji} must be small compared to B_{ij} unless ν is large. Spontaneous emission is therefore most often observed directly in the visible and ultra-violet regions. Its observation also requires non-equilibrium conditions such as discharges. In contrast the induced emission is unimportant except when ν is small, as otherwise $n_j \ll n_i$ and the number of molecules under equilibrium conditions which are capable of emission is negligible.

Transition moments and selection rules

Although the B_{ij} are molecular properties, the more usual quantities for molecular interpretation are the transition moments p_{ij}. These are related to the B_{ij} by

$$B_{ij} = p_{ij}^2(6\varepsilon_0\hbar^2)^{-1}$$

for molecules with random orientation. It should be noticed that B_{ij} is the essential experimental quantity and is always positive. p_{ij} is a vector and its direction is not determined by a measurement of B_{ij}.

Evaluation of p_{ij} in terms of the molecular wave function is an important problem in the theoretical treatment of any system. In common terminology

$$p_{ij} = \int \Psi_i p \Psi_j \, d\tau$$

where Ψ_i and Ψ_j are the wave functions for states i and j and where p is the dipole moment vector operator and the integration is to be taken over all coordinates including spin coordinates if relevant. In the Dirac terminology, which is finding increasing favour, one would write $p_{ij} = \langle i|p|j\rangle$. In Cartesian coordinates the x component of p is given by $\Sigma_k Q_k x_k$ where Q_k is the charge on the kth particle whose x coordinate is x_k. p_{ij} is a vector having x, y and z components.

In the approximation in which energies are the sums of energies of different types, vibrational, rotational, etc., the wave functions are product wave functions and the transition moments are products of integrals over different coordinates. It is only necessary for one of these factors to be zero for p_{ij} to be zero. The corresponding $i \leftrightarrow j$ transitions have zero intensity and are said to be *forbidden* or *inactive* in contrast to those with finite intensity which are called *allowed, permitted* or *active* transitions. It must be emphasized that forbidden transitions may have zero moment only in the approximation of separable wave functions. It is not unusual to observe forbidden transitions experimentally as very weak transitions and to calculate the transition moment theoretically from more exact wave functions.

Many of the integrals which arise in the transition moment occur repeatedly in different molecules. Those integrals which give rise to zero transition moments have become well known and their nature codified in a number of simple rules, the *selection rules.* These are remembered by most active workers in the field and are frequently quoted in support of particular spectral interpretations. There are, however, rather too many rules to be remembered by those who do not require them regularly. A typical rule would be $\Delta J = \pm 1$ which is to be interpreted to mean that the rotational quantum number, J, changes by one unit during an allowed transition. All other changes of J are forbidden for the class of transitions to which the selection rule applies. It is not necessary to specify whether absorption or emission is envisaged since $p_{ij} = p_{ji}$ and the one selection rule covers both cases.

As a simplified example of the derivation of such rules consider a linear diatomic molecule with unnormalized rotational wave functions $\psi_{J=0} = 1$; $\psi_{J=1} = \cos\theta$; $\psi_{J=2} = (3\cos^2\theta - 1)$; a vibrational wave function ψ_v and a dipole moment function $p\cos\theta$. For the $J = 1 \leftrightarrow 0$ transitions the transition moment is given by

$$\begin{aligned} p_{0,1} &= \int \psi_v \psi_{J=0} (p\cos\theta) \psi_v \psi_{J=1}\, \mathrm{d}\tau \\ &= \int \psi_v^2 p\, \mathrm{d}r \int_0^\pi \psi_{J=0} \cos\theta\, \psi_{J=1} \sin\theta\, \mathrm{d}\theta \\ &= p \int_0^\pi \cos^2\theta \sin\theta\, \mathrm{d}\theta = p \Big| -1/3 \cos^3\theta \Big|_0^\pi = 2p/3. \end{aligned}$$

In this derivation $\mathrm{d}\tau$ is the volume element and is taken to be $\mathrm{d}r \sin\theta\, \mathrm{d}\theta$.

For the $J = 2 \leftrightarrow 0$ transition the second integral is

$$\int_0^\pi (3\cos^2\theta - 1) \cos\theta \sin\theta\, \mathrm{d}\theta = 0$$

and the transition is forbidden. In the above derivation $\int \psi_v^2 p\, \mathrm{d}r$ has been put equal to p as ψ_v is taken to be normalized and p is assumed independent of r. If this assumption had not been made and if v had changed during the transition the integral would have been written $\int \psi_{v'} p \psi_{v''}\, \mathrm{d}r$ and would have had a different numerical value. However, the second integral in θ would have been unchanged as would the selection rule $1 \leftrightarrow 0$; $2 \nleftrightarrow 0$ for J. This rule is therefore the same for the pure rotational spectrum in the microwave region, when v is unaltered, and the rotational structure of the vibrational spectrum in the infra-red region when v changes. The symbol $\nleftrightarrow$ is a convenient shorthand for saying that the transition between the specified states is forbidden. Despite the simplification of this problem by omission of the azimuthal angle, it still provides an example of the more general rule $\Delta J = \pm 1$ for linear molecules.

It is to be understood that p is the dipole moment as a function of all coordinates when inserted into an integral. Even though the average dipole may be zero for the ground state it will be non-zero for distorted species and the function p will be non-zero. Molecules with zero dipole moments may still have vibrational transitions and monatomic atoms have allowed electronic transitions. They will not have active pure rotational transitions since the factor $\int \psi_v^2 p \, dr$ in the transition moment will always be zero.

In the previous paragraph p has represented the electric dipole, but it may also refer to the magnetic dipole. If the integral is zero for the electric dipole, but non-zero for the magnetic component, the transitions are commonly weaker than electric dipole allowed transitions by a factor of about 10^{-5}. Such transitions are said to be magnetic dipole transitions.

If the molecules are held rigidly and plane polarized radiation is used, p_{ij} must have a component parallel to the electric field of the radiation. If the field and moment vectors are inclined at an angle α the transition intensity is proportional to $\cos^2 \alpha$. For magnetic transitions it is the angle with the magnetic field vector of the radiation which must be used. With plane polarized radiation and a suitable single crystal of known structure the vector direction of p_{ij} can be experimentally located in the molecule. Moreover, the rotational selection rules for vibrational and electronic transitions depend on the direction of the transition moment vectors with respect to the axes of inertia, and so detailed interpretation of rotational fine structure may lead to similar information.

Line widths

The radiation absorbed or emitted is only monochromatic to the extent that the molecular energy levels themselves are sharply defined. If the energy gap has an uncertainty ΔE, then ν has an uncertainty $\Delta\nu \sim \Delta E/h$. Experimentally this appears as a line half-width $\Delta\nu$ at half the maximum intensity. $(2\pi\Delta\nu)^{-1}$ is the time associated with this uncertainty and may, in a sense, be considered the time over which the absorption or emission takes place.

The ultimate limit on the smallness of $\Delta\nu$ is controlled by the probability of spontaneous emission which limits the average lifetime of the excited state to A_{ji}^{-1} and the line width to A_{ji} rad s^{-1}. Since A_{ji} varies as ν^3 it is more important at high frequencies.

Another limiting factor is the rate at which transitions are taking place in the radiation field. If the measuring field is strong, a line broadening results which is associated with the power saturation. Even with weak measuring fields the radiation associated with the temperature of the sample and its container can cause a line broadening, $\Delta\nu = (4\pi kTA_{ji}/h\nu)$.

Except under rather special conditions the line width is in practice controlled by collisions which destroy the transition process. $(2\pi\Delta\nu)$ is then the collision frequency; under certain conditions this is proportional to gas pressure, especially if wall collisions are unimportant. $\Delta\nu$ is not necessarily the same for all transitions in one molecule. If attempts are made to interpret the collision frequencies on a gas kinetic model this leads to a collision diameter which varies with the transition concerned. For most cases the collision diameter is greater than would be expected on a hard-sphere model and may reach 15 Å for such simple molecules as ammonia. For liquids the molecules are in almost permanent collision and an averaging occurs so that liquid lines may be narrower than those of a high pressure gas.

If there is a relative velocity between the observing instruments and the molecules a Doppler effect results. The apparent frequency is given by $\nu(1+u/c)$ where $+u$ is the velocity of approach of the source to the detector. Such shifts are common in astronomical work and may be used to measure stellar velocities. In hot gases there is a distribution of molecular velocities and their directions so that a Doppler broadening can result. The line width is given by $\Delta\nu = [2\nu^2 kT(\ln 2)/mc^2]^{1/2}$ where m is the mass of the molecule.

When more than one form of line broadening is present the total line width can be obtained as the sum, i.e.

$$\Delta\nu = \Delta\nu_1 + \Delta\nu_2 + \Delta\nu_3 \ldots$$

where the subscripts refer to the different mechanisms. The majority of spectral lines are *Lorentzian* in shape which means that the absorption or emission of a single line of resonance frequency, ν_0, as a function of observing frequency, ν, is given by

$$\frac{C}{(\nu - \nu_0)^2 + (\Delta\nu)^2}$$

where C contains only frequency independent quantities, including the square of the transition moment for the line in question.

Associated with the absorption of radiation is a related change in the real part of the refractive index. Figure 2.2 shows the form of this relationship graphically; it should be noticed that the refractive index rises on the low-frequency side, reaches a maximum, passes through its mean value at the centre of the absorption, reaches a minimum, and finally approaches a horizontal asymptote on the high frequency side. This asymptote is lower than the high frequency asymptote by a small difference, Δn, which is proportional to the area of the absorption curve. This is a special case of the *Kramers-Kronig relations* which connect the absorption at any frequency to the whole of the refractive index curve and

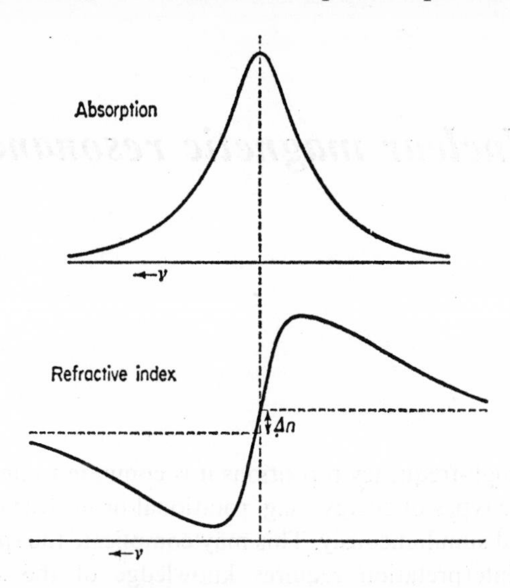

2.2 Relationship of refractive index with frequency showing anomalous dispersion features near an absorption line.

likewise relate the refractive index at any frequency to the whole of the absorption curve. For more than one absorption line the effects must be summed.

Nuclear magnetic resonance

3

In studying high-frequency transitions it is common to find that changes in the smaller types of energy—e.g. rotational or nuclear orientational—have occurred simultaneously. This may complicate the spectra observed and, since interpretation requires knowledge of the smaller energy intervals, it is most appropriate to discuss these first. Consequently two forms of nuclear orientation phenomena are discussed in this chapter and the next.

Nuclear properties

To understand the processes involved some knowledge of nuclear properties is required. Nuclei are well known to possess a positive charge Ze, where Z is the atomic number which distinguishes the element and e the charge on the proton. They also have a mass m which can vary from one isotope to another of the same element. Nuclei may also possess spin, a magnetic dipole moment, an electric quadrupole moment and occasionally higher moments. Intrinsic nuclear angular momenta are quantized and may be expressed as $I\hbar$ where I is an integer or half-integer and is called the spin quantum number. More briefly a nucleus for which $I=3/2$ is said to have a spin of 3/2. I may be different for different isotopes. There is a restriction that for nuclei of even mass number I must be an integer or zero whereas for nuclei of odd mass number I must be a half-integer. Most heavier stable nuclei of even mass have zero spin. Table 3.1 indicates some nuclear properties, including spin, of some common nuclei.

If a nucleus has zero spin, all its moments are zero and no nuclear orientational effects can arise. If the spin is 1/2 or greater the nucleus possesses a magnetic moment, μ; in this it resembles any other rotating charge. The nucleus may be thought of as incorporating a little magnet whose direction is fixed rigidly parallel to the spin axis. A negative moment implies that the moment vector is antiparallel to the spin vector. The unit used to express nuclear moments is the nuclear magneton which is $e\hbar/2m_p$ where m_p is the mass of the proton. The symbol is μ_N and the value 5×10^{-27} m^2 A. Nuclei of spin 1 or greater possess an electrical quadrupole moment, as is discussed more fully in chapter 4.

Table 3.1 *Some nuclear properties*

NUCLEUS	SPIN I	MAGNETIC MOMENT (μ/μ_N)	RESONANCE FREQUENCY $(\nu B^{-1}/\text{MHz T}^{-1})$	QUADRUPOLE MOMENT $(Q/10^{-28}\text{ m}^2)$
H	1/2	2·79	42·6	–
D	1	0·86	6·5	0·0028
^{4}He	0	–	–	–
^{12}C	0	–	–	–
^{13}C	1/2	0·70	10·7	–
^{14}N	1	0·40	3·1	0·02
^{16}O	0	–	–	–
^{19}F	1/2	2·63	40·1	–
^{23}Na	3/2	2·22	11·3	0·1
^{31}P	1/2	1·13	17·2	–
^{32}S	0	–	–	–
^{35}Cl	3/2	0·82	4·2	−0·08
^{37}Cl	3/2	0·68	3·5	−0·06
^{39}K	3/2	0·39	2·0	0·07
^{79}Br	3/2	2·10	10·7	0·33
^{81}Br	3/2	2·26	11·5	0·28
^{127}I	5/2	2·79	8·5	−0·75

Basic principles of nuclear magnetic resonance

The angular momentum vector of a nucleus can take up $(2I+1)$ directions in space and these directions are most often characterized by the resolved angular momentum along a specified direction. The resolved momentum is quantized and represented by M_I. Strictly the resolved momentum is $M_I\hbar$ and M_I may have the values $I, I-1, I-2, \ldots -I+1, -I$. For the common case of $I=1/2$, $M_I=+1/2$ or $-1/2$ and there are only two states of nuclear orientation. Transitions between these states are magnetic dipole allowed and can be induced by an alternating magnetic field of the correct orientation and frequency. In free molecules, however, the energy difference between the two states is so small as to be effectively unobservable. But if a magnetic field is present there is an additional energy parameter related to the energy of the nuclear moment in this field. This is analogous to the classical energy of a compass needle at an angle θ to a magnetic field which has energy $-\mu B\cos\theta$. The quantum mechanical equivalent of this energy is $-(\mu/I)BM_I$ where B is the magnetic induction and M_I is the quantum number of the projection of the spin on the magnetic field direction.[1] The energy may alternatively be written in terms of the scalar product (dot product) of the field and the spin vector operator, $\boldsymbol{I}$, namely as $-(\mu/I)\boldsymbol{B}\cdot\boldsymbol{I}$.

For the transition $M_I = -1/2 \leftrightarrow +1/2$ the Bohr frequency rule gives

$$h\nu = -(\mu/I)B(-\tfrac{1}{2}-\tfrac{1}{2})$$

or $\nu = (\mu/Ih)B.$

This equation may also be written in terms of the *magnetogyric ratio*, γ, where $\gamma = \mu/I\hbar$, so that it becomes

$$\omega = 2\pi\nu = \gamma B \text{ rad s}^{-1}.$$

The same resonance condition holds for any nuclear resonance transition for which $\Delta M_I = \pm 1$ which is the selection rule for such processes. Table 3.1 contains a column which gives the frequency in an induction[2] of 1 Tesla for the nuclei selected. This indicates that, for hydrogen nuclei for example, resonance would occur at 42·6 MHz in a field of 1 T, but for resonance at 100 MHz a field of 2·35 T would be required as indicated in Figure 3.1. The analogue of the resonance frequency in classical mechanics is the gyroscopic precession frequency of a spinning magnet about a magnetic field. Such nuclear resonance was first observed in 1946 by Purcell and Pound and by Bloch.

For most purposes the transition is observed at several MHz in inductions of several Tesla but it has proved possible to obtain signals for hydrogen nuclei in the earth's magnetic field. Frequencies may be measured with considerable accuracy and for well-known nuclei the effect may be used to measure B. A number of portable magnetometers embodying nuclear magnetic resonance devices are available for measuring laboratory magnets and for magnetic survey work in geophysics and archaeology.

The transitions with $\Delta M_I = \pm 1$ are magnetically allowed provided the radiofrequency field causing them is perpendicular to the static field. The transition moment is determined by the nuclear moment and consequently it is independent of the chemical compound which contains the

[1] The equation corresponds to the normal definition and μ then corresponds to the magnetic moment quoted in tabulations. However, even when $M_I = I$ the nucleus is aligned at an angle $\cos^{-1}[I/(I+1)]^{1/2}$ to the field, so that were classical laws applicable the total magnetic moment would be $\mu[(I+1)/I]^{1/2}$. Under quantum laws this larger moment can never be observed.

[2] The SI unit of magnetic induction, B, is the tesla = kg s^{-2} A^{-1}. The field, H, should not be used in these algebraic formulae as was the custom with emu. Since a magnetic induction implies a magnetic field the name field is acceptable in imprecise contexts.

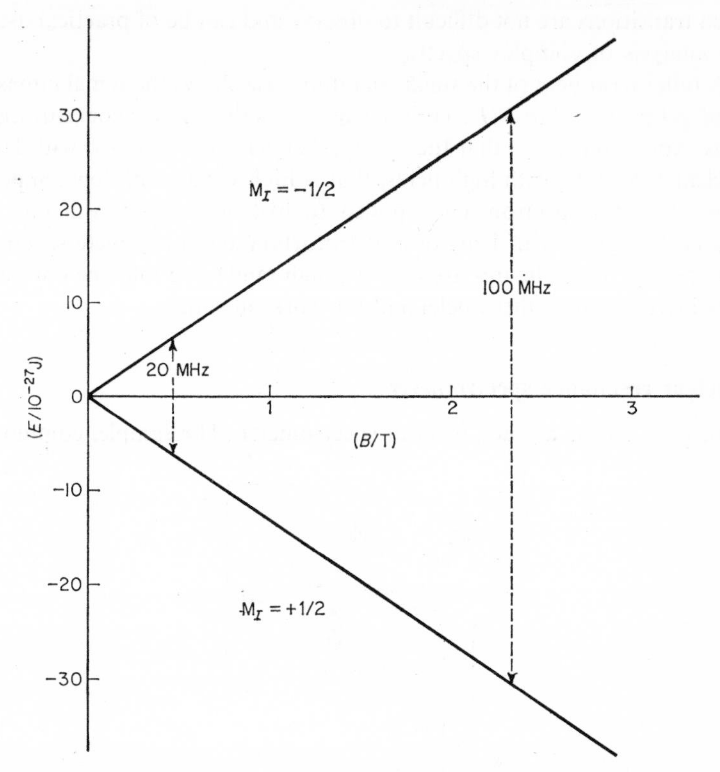

3.1 Variation of the energy and transition frequency with magnetic induction for a single hydrogen nucleus.

nucleus. This means that for quantitative analysis it may be assumed that the area of the signal is proportional to the number of nuclei present. Since, as explained below, the same type of nucleus in two different chemical environments may give separated signals, the number of nuclei in each environment may be measured. Also isomeric analysis does not require that the separate isomers be available in the pure state, thus enabling a direct analysis of interconverting isomers as in keto-enol tautomerism.

If more than one nucleus is present the $\Delta M = \pm 1$, single quantum, selection rule may break down in the presence of strong radiofrequency fields and $\Delta M = \pm 2$, or higher, transitions can be observed. The $\Delta M = \pm 2$ case is called a *double quantum* transition and its probability is proportional to the square of the radiofrequency radiation density. The Bohr energy condition is now

$$E' - E'' = 2h\nu.$$

Such transitions are not difficult to observe and can be of practical use in the analysis of complex spectra.

A fuller treatment of the single quantum case shows the signal intensity to be proportional to $N(I+1)\mu^2\nu/T$ where N is the volume concentration. This expression shows that the strongest signals are obtained with large nuclear moments, with high fields, that is high ν, and with low temperatures. Modern spectrometers applied to hydrogen nuclei will give an observable signal with 1 mg of a hydrocarbon and a complete spectrum with 50 mg. These figures are only a rough guide and must be considerably increased for other nuclei and for work on solids.

Nuclear resonance spectrometer

Figure 3.2 shows a crude form of spectrometer. The sample, commonly

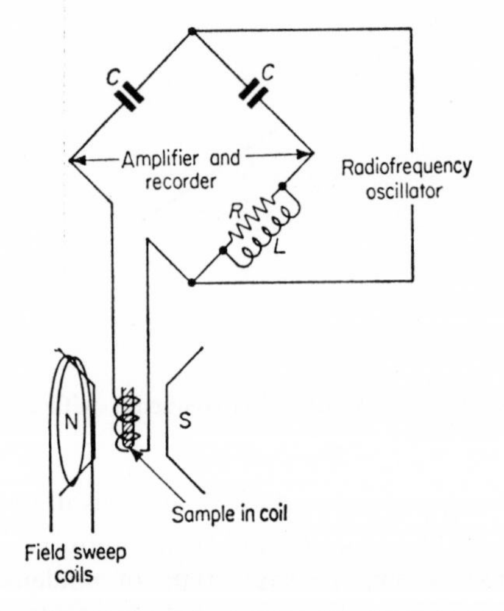

3.2 Elementary form of nuclear magnetic resonance spectrometer.

in a small glass tube, is placed in a strong magnetic field and located inside the coil of an inductance whose interior volume it fills as fully as possible. This coil is one arm of a balanced radiofrequency bridge which is fed from a radiofrequency oscillator at the resonance frequency. The radiofrequency field generated in the coil is partly absorbed by the nuclear system and the loss of energy appears as an apparent increase of resistance in the coil which leads to an unbalance of the bridge. A signal appears across the detector section of the bridge which can be amplified and recorded. The most convenient way to display the signal is to vary the magnetic field slightly as a function of time so that it crosses the resonance value. A single resonance then appears as in Figure 3.3(a) where the regular convention has been followed in showing the magnetic field increasing to the right and the signal plotted upwards. In some cases the signal is followed by a damped oscillation or 'wiggles' as in Figure 3.3(b). This extraneous feature may be eliminated by using slower sweep rates, but this is not always convenient. The wiggles can be ignored for most purposes. Other forms of spectrometer change the frequency, when convention requires that signals should be displayed with lower frequencies on the right.

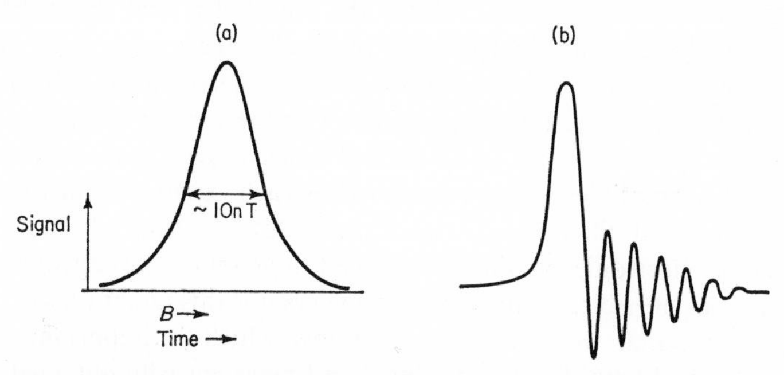

3.3 High resolution nuclear magnetic resonance absorption; (a) Ideal slow passage, (b) As commonly observed showing 'wiggles'.

In view of the numerous chemical applications to be described below, nuclear resonance spectrometers are to be found in many large research laboratories and instruments made commercially have reached a high state of sophistication.

Some of the difficulties are considerable especially when the highest resolving power of 10^9 or greater is required. In all branches of spectroscopy the *resolving power* of an instrument is a measure of the ability of a spectrometer to distinguish closely spaced lines. Thus an instrument of high resolving power will show two closely spaced spectral features correctly whereas in a low-power instrument these will be merged together. If $\Delta\nu$ is the frequency separation of the closest lines which can be clearly seen to be two lines the resolving power is $\nu/\Delta\nu$ there ν is the mean frequency of the lines. $\Delta\nu$ is called the *resolution* which would be 0·5 Hz for an instrument working at 50 MHz with a resolving power of 10^8. Under some circumstances the resolution is limited by the line width of the sample and not by the nature of the spectrometer. To produce an oscillator stable to 1 in 10^8 is not exceptionally difficult with quartz crystal oscillators. Magnetic field stability to this precision is more trouble and may involve mains stabilization, current stabilization and field stabilization with negative feedback as well as thermostatic control of the magnet and its cooling water. If permanent magnets are used the power supply difficulties are reduced, but the highest fields are not easily reached and thermostatic control of the whole magnet to $\pm 10^{-3}$ K is required. Even more difficult still is the achievement of a constant field over the entire sample volume which may be 5 mm in diameter by 1 cm long. Inhomogeneities over the sample means that the resonance condition will not be reached simultaneously for the whole sample and the observed line will be broadened. The use of large magnets up to a Mg in mass with carefully polished and aligned pole faces improves the homogeneity. Further improvement follows the use of electrical shims, which are essentially carefully shaped loops of wire on the magnet face; these carry small currents which generate local magnetic fields which are arranged to cancel the inhomogeneities. For the highest resolution the sample is rotated by an air jet so that all molecules are made to visit all parts of the accessible sample volume. They each experience the average field and so local differences between molecules are effectively removed.

The main electronic circuits are designed to present as large a *signal-to-noise* ratio as possible. As in all spectrometers it is this quantity, the ratio of genuine signal to random fluctuations, which is important. Any degree of equal amplification of signal and noise is easily obtained, but provided the signal is sufficient to operate the recording system chosen, increased amplification is valueless. There is always one way to reduce noise, namely to average the random fluctuations over a long period of time. The noise amplitude is inversely proportional to the square root of the time constant of the system and consequently greater sensitivity is obtained by displaying the spectrum in 10 minutes on a damped pen recorder than by presenting the spectrum in 10^{-2} seconds on an oscilloscope screen. It is also possible to observe a spectrum many times over and to add the observations together; in this way the signals are added and the noise features are partially self-cancelling. Again the improvement in signal-to-noise generally varies as the square root of the time taken, which means that the signal-to-noise ratio is proportional to the square root of the number of times the spectrum is observed. For the special case of magnetic resonance the improvement is rather greater because of the diminution of power saturation effects, and devices for such averaging, colloquially known as CATs, are being used increasingly often.

Subsidiary electronic circuits are usually provided for changing the magnetic field at varying rates over ranges of different lengths, for measuring the separation between two resonance signals, for changing radiofrequency to suit different nuclei, varying the radiofrequency power in the coil, etc. Provision may also be made for presenting the integrated signal for quantitative analysis or the derivative of the signal for broad line work on solids.

Chemical shifts

The nuclear resonance equation, $\nu = (\mu/Ih)B$, would suggest that all nuclei of the same isotope in the same static magnetic field would have the same resonance frequency. To a very large extent this is true, but precise measurements reveal that differences occur which are of the order of 1 in 10^6 for hydrogen and even larger for other nuclei. The exact resonance frequency depends on the chemical structure of the molecule or grouping containing the nucleus. These differences are called chemical shifts and denoted by a chemical shielding factor, δ, such that

$$\nu = (\mu/Ih)B(1+\delta).$$

This small correction factor arises from the diamagnetic character of the molecules. The effect arises from the same essential cause as static diamagnetism, that is from the circulation of the electrons round the magnetic field in so far as the molecular structure will allow this. The moving charges then set up a magnetic field which opposes the static field. Since these fields arise from internal features of the molecule they are not uniform over small distances and their magnitude and direction vary from one part of the molecule to another so that δ varies for individual nuclei of the same kind. Even for one nucleus δ varies with the direction of the magnetic field if the molecule is kept in a fixed orientation. In mobile liquids or gases individual molecules are rotating rapidly and only average shifts are effective.

Although it would be most logical to express chemical shifts with respect to a bare nucleus, this proves to be both difficult and inaccurate. An artificial origin of zero shift is therefore defined. Sometimes a standard substance in an external tube is used as a reference with all samples. However, there is a correction to be made for the shape of the tube and the bulk diamagnetism of both sample and reference material and the necessary data is not always available. More convenient is an internal standard a few per cent of which is added to the sample tube. For hydrogen resonances tetramethyl silane is a very common standard since it is chemically inert and gives a single resonance signal which appears at the high field edge of the spectrum and seldom obscures the spectrum of interest. One scale based on tetramethyl silane is often called the τ scale. The definition of τ is given by

$$\tau = 10{\cdot}000 - 10^6(B_{\mathrm{TMS}} - B)/B_{\mathrm{TMS}}.$$

In this expression B_{TMS} is the resonance field for tetramethyl silane, $(CH_3)_4Si$, and B the field for the nucleus in question at the same frequency. To a good approximation chemical shifts so expressed are independent of magnetic field and of solvent, providing specific solute-solvent interactions are absent. Shifts are occasionally quoted in Hz but in this form they are field dependent.

Since the major effect comes from electrons in the vicinity of the nuclei, the chemical shifts of any grouping are relatively independent of the remainder of the molecule. Accordingly it is possible to construct a table of typical chemical shifts for any grouping. Table 3.2 gives the τ

Table 3.2 *Abbreviated chemical shift table for compounds CH_3X*

NATURE OF X	τ
—$Si(CH_3)_3$	10·0
—R	9·1
—cyclic saturated hydrocarbon	9·1
—$CR{=}CR_2$	8·3
—COR	7·9
—I	7·84
—NR_2	7·8
—aromatic hydrocarbon	7·7
—Br	7·32
—Cl	6·95
—OR	6·8
—N^+R_3	6·7
—O–COR	6·3
—F	5·74
—NO_2	5·72

R = aliphatic group

values of a number of types of methyl group compounds. Where only a type of substituent is indicated the shifts may vary by 0·2 between compounds, but for restricted classes the range may be even narrower. Such narrow ranges make the observation of chemical shifts for compounds of uncertain structure an important tool in organic chemistry. A large range of chemical shift tables are available for such analysis of unknown compounds and the literature on this topic is growing rapidly. Nuclei in equivalent chemical positions, such as the three hydrogens of one methyl group, always have identical chemical shifts.

Theoretical chemists have had some success in calculating chemical shifts from wave functions and most of the qualitative features of observed chemical shifts are understood. High diamagnetic shielding, and related large τ values, occur when the nucleus is embedded in a cloud of electrons. Conversely electronegative groups attract electrons away from the nucleus and low shielding results; this effect is exemplified in nitromethane and methyl fluoride which have exceptional low τ values as can be seen in Table 3.2. A more subtle effect is due to mobile electrons adjacent to the nucleus in question. The large associated diamagnetism can be thought of as repelling the magnetic field which is therefore increased at the position of the nucleus if this is beside the diamagnetic centre, Figure 3.4(a), and reduced if it is oriented as in Figure 3.4(b). If

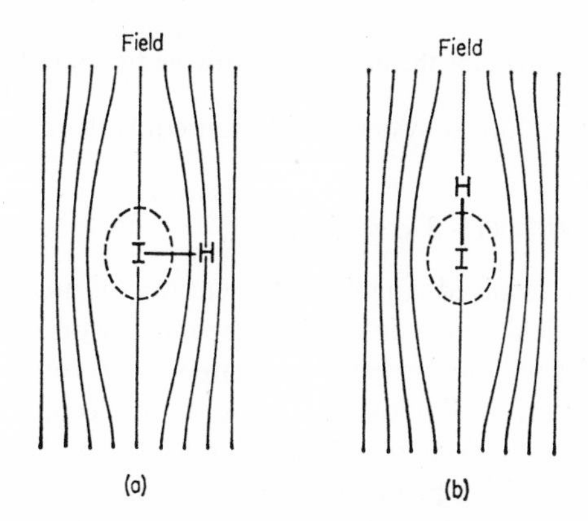

3.4 Effect of diamagnetic material (shown as a dotted sphere) on the effective magnetic field at an adjacent hydrogen nucleus.
(a) if the bond to the hydrogen is perpendicular to the field when poor shielding or de-shielding results; (b) the bond is parallel to the field and the shielding is high. The sphere might be an iodine atom, which, being diamagnetic, repels the field.

the diamagnetism is isotropic the effect averages to zero for rotating molecules, but otherwise there will be a non-zero average. For peripheral atoms on an aromatic ring the average effect is deshielding. This accords with the lower τ value of a methyl group attached to an aromatic residue compared to an aliphatic residue as shown in Table 3.2. Effects of this nature can arise from solvent molecules; solvent effects, if interpreted with care, can give information about hydrogen bonding to the solvent and other specific interactions.

Spin-spin coupling

Another reason why the magnetic field at the position of a particular nucleus is not the same as that applied by the magnet is because there are local fields originating in the magnetic moment of other nuclei in the vicinity. In this case it is more convenient to examine the relevant energy levels than the magnetic fields themselves.

Classically the energy of interaction of two small, parallel magnets of moments μ_1 and μ_2 at a distance r and inclined at an angle θ to the line joining their centres (see Figure 3.5) is given by $\mu_0\mu_1\mu_2(1 - 3\cos^2\theta)(4\pi r^3)^{-1}$.

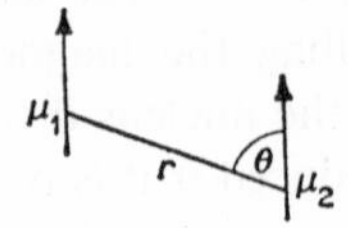

3.5 Coordinate system for the interaction between two parallel point dipoles.

For quantum mechanical systems with non-equivalent nuclei in a strong field this energy is

$$(\mu_0\mu_1\mu_2/4\pi r^3 I_1 I_2)(1-3\cos^2\theta)M_1M_2.$$

Here M_1 is the resolved angular momentum of nucleus 1 along the strong magnetic field which determines the nuclear orientations; this and related quantities will be further abbreviated as M_1, etc., for the remainder of this chapter. In liquids and gases the molecules are rotating or 'tumbling' and in random translational motion so that θ takes on all values from 0 to π with a probability proportional to $\sin\theta\, d\theta$. The energy of interaction averaged over this motion is proportional to

$$\int_0^\pi (1-3\cos^2\theta)\sin\theta\, d\theta = 0,$$

and the average interaction vanishes. In solids (see p. 40) this averaging does not apply and the term is of great importance.

However, an energy term of the same general form, $hJ\, M_1M_2$, must be introduced to explain all features of high resolution spectra. This energy is seen to be proportional to the resolved spin momentum of each nucleus and from this property it is called the spin-spin interaction. For equivalent nuclei and low magnetic fields the interaction energy should be written in its more exact form $hJ\, \mathbf{I}_1 \cdot \mathbf{I}_2$ where $\mathbf{I}_1$ and $\mathbf{I}_2$ are the spin vector operators and the dot represents the scalar product.

A molecular fragment of the form ^{13}C–H in which a hydrogen atom is directly bonded to the 1 per cent abundant spin 1/2 nucleus of carbon-13 makes a convenient example to illustrate the origin and effect of this spin-spin coupling. Each nuclear moment interacts, as explained more fully in chapter 5, with electron magnetic moments. Specially important is the interaction with the two electrons which form the bond and each nucleus attracts preferentially that electron whose magnetic moment is parallel to that of the nucleus. By virtue of the Pauli exclusion principle the two electrons of the bond are of opposite spin to each other. If the H and ^{13}C moments are anti-parallel to each other, that is the product $M_H M_C$ is negative, each nucleus attracts a different electron and these will each favour, fractionally, a different end of the electron pair bond with a consequent small decrease in energy. In contrast, if the H and

^{13}C moments are parallel they compete and try to attract the same electron of the pair and this leads to a higher energy. J represents the size of these energy terms and may amount to 120 Hz for the ^{13}C–H bond. For indirect bonding the values are much lower and the detailed explanations more involved. In organic molecules hydrogen-hydrogen couplings are less than 20 Hz and may even be slightly negative. Couplings involving atoms of higher atomic number, such as ^{19}F, ^{31}P and ^{117}Sn, tend to be rather larger. Table 3.3 shows a further selection of values.

Table 3.3 *Typical spin-spin coupling constants (J/Hz)*

DIRECT		INDIRECT H . . . H		
H–H	+280	CH–CH		+ 5
H–^{11}B	+ 90	>CH_2		−10
H–^{13}C	+120			
H–^{14}N	+ 50	C=CH_2		0
H–^{19}F	+615	CH=CH	cis	+10
		,,	trans	+18
		Aromatic	ortho	+ 6
		,,	meta	+ 2
		,,	para	+ 1

The effect of such a coupling on the spectrum can be understood from Figure 3.6 which shows the energy diagram. There are four possible spin states corresponding to the values of M_C and M_H indicated on the left. The main energy terms, $-(\mu/I)B_0M$ are written as $-\nu M$ in frequency units where ν is the nuclear resonance frequency in the magnetic field. If this were 1 T, ν_C would be 10·7 MHz and ν_H 42·6 MHz. The energy contribution $hJ\,M_HM_C$ provides a small correction and the values are given on the right of Figure 3.6. The selection rules are $\Delta M_H=1$ and $\Delta M_C=0$ or $\Delta M_H=0$ and $\Delta M_C=1$ which lead to the four allowed transitions indicated. (i) and (ii) represent two ^{13}C transitions at $\nu_C-J/2$ and $\nu_C+J/2$, while (iii) and (iv) are two hydrogen transitions at $\nu_H-J/2$ and $\nu_H+J/2$. Each spectral region consists of two transition frequencies separated by J Hz. Since (i) and (ii) and likewise (iii) and (iv) are of equal intensity it is not generally known which is which. That is the magnitude, but not the sign, of J is obtained. When two or more different J occur in one molecule it may be possible to determine their relative signs but the absolute signs are missing. The absolute signs are usually inferred from discussions based on the electronic structure or by analogy with similar compounds. Experimental methods are based on spoiling the perfection of the 'tumbling' motion in the liquid by strong electric fields or the use of liquid crystals; these methods are difficult to apply and in some cases the sign of J remains unknown.

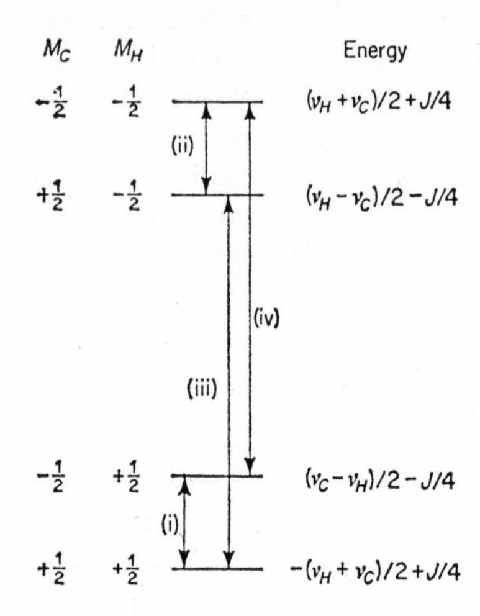

3.6 Energy diagram for a ^{13}C–H fragment in a magnetic field.

If the two nuclei are fully equivalent to each other in their environment, as in F_2, slightly different considerations apply. The resolved spin of each nucleus along the magnetic field direction is no longer a satisfactory quantum number for describing the system's behaviour. Instead the individual nuclear spins are compounded to give a total spin I for the set of equivalent nuclei and then there are the $(2I+1)$ values of resolved spin, M. For F_2 the total nuclear spin may be 1 or 0, corresponding to parallel and anti-parallel nuclei. Figure 3.7 shows the energy level diagram (not to scale) and the transitions. This spin coupling energy, $-3J/4$, for the anti-parallel states is derived from the full form of the coupling $J\,\mathbf{I}_1\,.\,\mathbf{I}_2$. The selection rules are $\Delta I=0$, $\Delta M=1$ and the allowed

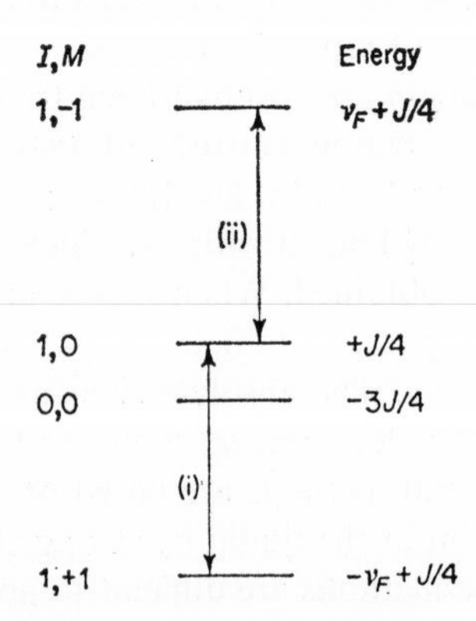

3.7 Energy diagram for F_2 in a magnetic field.

transitions, (i) and (ii), are independent of J and are exactly superposed at the frequency, ν_F, to be expected if $J=0$.

If a nucleus couples to more than one other nucleus, its resonance is further split. Equivalent nuclei have the same coupling to any other nucleus and characteristic patterns arise. A symmetrical pattern with 1 : 3 : 3 : 1 intensity ratios suggests a methyl group as in CH_3F. The four lines of the fluorine resonance arise, essentially, from the eight possible hydrogen spin arrangements which may be designated $(+++)$: $(-++)$, $(+-+)$, $(++-)$: $(+--)$, $(-+-)$, $(--+)$: $(---)$. The coupling energies are $M_F\,3J/2$: $M_F\,J/2$, $M_F\,J/2$, $M_F\,J/2$: $-M_F\,J/2$, $-M_F\,J/2$, $-M_F\,J/2$: $-M_F\,3J/2$ respectively and since $\Delta M_F=1$ the four lines of separation J and intensities 1 : 3 : 3 : 1 result. In the hydrogen resonance spectrum of CH_3F the H . . . H coupling does not influence the result and two lines at $\nu_H \pm J/2$ are found as in HF. It is because all energies are $\ll kT$ for any liquid that all spin orientations are equally probable according to the Boltzmann distribution.

For the same nucleus in inequivalent positions the simple theory, based on the example of ^{13}C–H, is appropriate provided the shift of frequency arising from the chemical shift is much greater than J. The higher the magnetic field, to which the chemical shift expressed in Hz is proportional, the more likely is this condition to be fulfilled. If J is of the same order of magnitude as the chemical shift a fuller theory must be employed. In this case extra, weak, lines may appear and the intensity ratios of the strong lines depart from integral values.

Examples of high resolution spectra

Interpretations of high resolution nuclear resonance spectra are based on the recognition of chemical shifts and spin-spin couplings and their effects on the spectra.

One simple example is the determination of the structure of the dimer of ketene. The medium resolution hydrogen resonance spectrum consists of two equal intensity single lines at $\tau = c.$ 5·4 and 6·1. The formula, $C_4H_4O_2$, shows hydrogen to be the only nucleus with a magnetic moment so that the two peaks must be each due to two hydrogens. Structures I to V have been proposed at various times for the structure of dimeric ketene; III, IV and V do not contain two pairs of hydrogen atoms and can be eliminated at once. A closer inspection of II shows all four hydrogens to be equivalent so that only one peak would be expected. Structure I would be expected to give two equal intensity lines and must be correct. Under the highest resolution the two lines show a small splitting due to the spin-spin coupling between the pairs of hydrogen atoms and to a

I II III IV

$CH_3COCH{=}C{=}O$

V

chemical shift between the ethylenic hydrogens which must be in the plane of the ring and not perfectly equivalent being cis and trans to the oxygen atom. (See also problem 9.)

Figure 3.8 shows the fluorine resonance of $CF_2{=}CHCl$. The two fluorines are non-equivalent being cis and trans to the hydrogen. There is a chemical shift of about 180 Hz at 40 Hz or 4·5 ppm between the centres of the four lines at low field and the two high field lines. This shift is larger than would be expected between hydrogens in analogous positions, but is typical for fluorine atoms. The further splitting must arise from spin-spin coupling and one $J = 45$ Hz is given immediately by the separation of the high field pair. This same interval separates the first and third

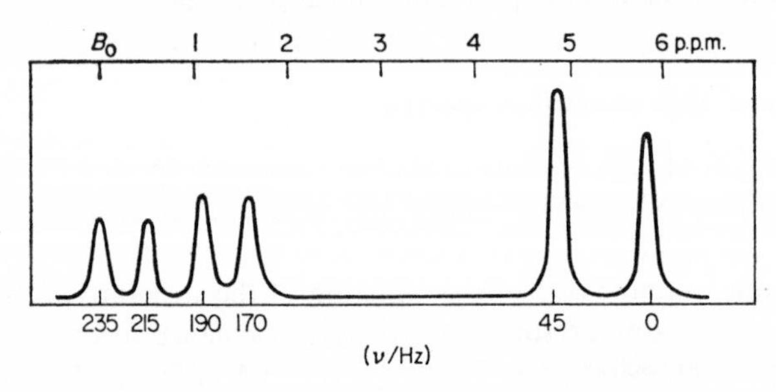

3.8 Fluorine resonance of $CF_2 = CHCl$ seen on a high resolution spectrometer at 40 MHz.

and likewise the second and fourth of the low field lines, which confirms the interpretation and also shows that this interval must be the fluorine-fluorine coupling. The remaining splitting, 20 Hz, between the first and second and likewise third and fourth low field lines must be between one of the fluorines and the hydrogen. The high field lines are not further split and the other hydrogen-flourine coupling must be less than the line width. Analogous compounds show $J_{trans} \gg J_{cis}$ so that the fluorine trans to the hydrogen must be responsible for the lower field pattern. These values $|J_{FF}| = 45$ Hz; $|J_{HF}|$ (trans) $= 20$ Hz; J_{HF}(cis) $= 0$ Hz; chemical shift 4·5 ppm (i.e. 180 Hz at 40 MHz) explain the position of all the lines quite closely but predict an intensity pattern 1 : 1 : 1 : 1 : 2 : 2 on the simplified theory. These ratios are obtained by requiring the total intensity of each fluorine to be the same and requiring the separate lines from a spin-spin splitting to be equal. Figure 3.8 shows the central lines to be stronger than the simple ratios suggest. This is very common and readily explained by an exact treatment for which it is no longer assumed that the chemical shift is much greater than the spin-spin couplings of similar nuclei. The absence of chlorine-fluorine spin-spin couplings is probably due to the relaxation of the chlorine nucleus by virtue of its quadrupole moment. Each chlorine nucleus is then equally likely even on a short time scale to be parallel or anti-parallel to the fluorine so that the average coupling is zero.

Chemical exchange reactions, internal rotations, etc.

Under favourable experimental conditions, the individual resonance lines may be less than 1 Hz in width. This implies (see chapter 2) that the individual energy state must remain undisturbed for longer than about 1 s for it to be specified with this precision. Many chemical processes occur faster than this and when operative these will make the energy states imprecise and broad lines result. Typical rapid reactions would be both the forward and reverse reactions in the equilibrium

$$CH_3COOH + H_2O \rightleftharpoons CH_3COO^- + H_3O^+.$$

If such reversible reactions are very rapid, the nucleus which exchanges will visit a large number of molecular sites and will experience exactly the average chemical shift appropriate to such sites. The average is precisely specified and sharp lines are again obtained. The effect of the exchange is therefore to make both the acid hydrogens and those of the water molecules and hydroxonium ions have the same chemical shift, since for an experiment which takes about a second the individual hydrogen nuclei cannot be classified in these separate ways. The actual chemical shift depends on the proportion of the groupings present and hence on concentration. The hydrogens of the methyl group do not partake of such exchange and give rise to a separate resonance signal. However, any one methyl group may have many different hydrogens on its adjacent carboxyl group and half of these will have their nuclear spin up and half will have it down. Averaged over the time of the experiment the net coupling energy between these different hydrogens and those of one methyl group will be zero and the spin-spin coupling between the methyl group and the acidic hydrogen is effectively removed by the exchange.

Not only intermolecular reactions but also intramolecular processes can produce this averaging effect. In cyclohexane at room temperature only one peak is observed, despite the existence of hydrogens in both axial and equatorial positions between which a chemical shift can normally be detected. The interconversion of one chair form into the other converts axial to equatorial hydrogens and *vice versa*, so that the time average positions of all hydrogens are the same.

For a unimolecular reaction the half life must be much less than the reciprocal of the separation of the peaks to be expected in the absence of the reaction if the spectrum is to collapse into one line. Variation of the reaction rate may reverse this inequality and the spectrum will change accordingly as in Figure 3.9. This shows the change of spectrum of

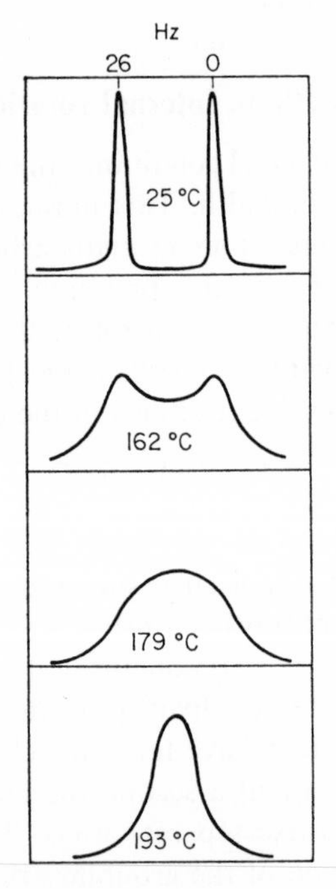

3.9 Variation of the hydrogen resonance of N-nitroso dimethylamine, $(CH_3)_2N{-}N{=}O$ with temperature at 40 MHz.

N-nitroso dimethylamine with the temperature. In a rigid form the two methyl groups are inequivalent being cis and trans to the oxygen atom but at high temperatures there is rapid rotation about the N–N bond and the spectrum coalesces to one line.

A variety of reactions and especially internal rotations can be studied in this way and half lives of the order of 10 ms determined from an analysis of line shapes. For faster reactions between non-equivalent species the value of an average chemical shift or spin-spin coupling may be used to measure the equilibrium constant, if the values of the pure forms can be estimated.

Double nuclear resonance

Another way in which spin-spin couplings can be annihilated is by the application of a strong radiofrequency field of the resonant frequency of one of the nuclei at the same time as the resonance of the other is being observed. The strong radiofrequency field induces both absorption and emission many times a second. The time average of the coupling of this decoupled nucleus to all others is thus zero. This double resonance technique produces a simplification of the spectrum and makes interpretation easier. It is easiest to apply when two different nuclei are involved, but special experimental arrangements can be made to work when the two nuclei are of the same type, so that the frequency difference is only that produced by the chemical shift. It is also possible to measure the exact frequency of the most effective strong radiofrequency field, which is the resonance frequency of the stirred nucleus. The detection sensitivity is that of the second or observed nucleus which may be much the greater if the first nucleus has only a small magnetic moment so that direct detection of its resonance is difficult. Sometimes too the signal from the first nucleus is lost in overlapping spectra. Double resonance may also be used to obtain the relative signs of different J with one nucleus in common.

Figure 3.10(a) shows the simple hydrogen resonance spectrum of $Al(BH_4)_3$ in which the coupling of ^{11}B and ^{27}Al of spins 3/2 and 5/2 respectively give rise to 24 overlapping lines which are observed as one broad maximum. Irradiation at the ^{27}Al resonance frequency gives a hydrogen resonance spectrum of Figure 3.10(b) which reveals the four peaks due to ^{11}B splitting and some much weaker lines due to splitting by ^{10}B which has a spin of 3 and is present in 19 per cent abundance.

These descriptions of double resonance are appropriate for strong radiofrequency fields, but more subtle experiments are possible if only medium strength fields are used. In particular if the radiofrequency field is of strength B_r such that $(\mu B_r/Ih)$ is about equal to the line width and is of a frequency exactly equal to that of an absorption line, then those other lines in the spectrum which share a common energy level with the irradiated line are split into two lines of equal strength separated by

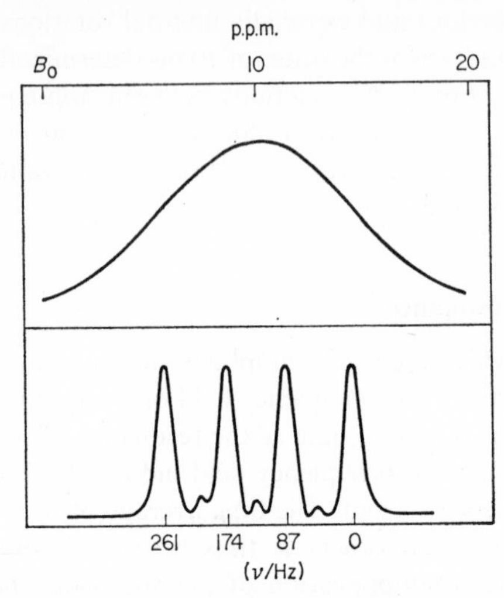

3.10 (a) Hydrogen resonance at 30 MHz of $Al(BH_4)_3$; (b) Same with ^{27}Al nuclei simultaneously stirred with a high intensity magnetic field at 7·8 MHz.

($\mu B_r/Ih$). The use of this technique, sometimes referred to as 'tickling', to indicate which lines do, and which do not, have common energy levels can be of great benefit in the analysis of complex spectra and in particular for the determination of the relative signs of spin-spin couplings.

Solids

For rigid solid systems each pair of nuclei are fixed with respect to each other and to the direction of the magnetic field. Consequently there is no averaging of the direction dependent spin-spin coupling. The energy term

$$(\mu_0\mu_1\mu_2/4\pi r^3 I_1 I_2)(1-3\cos^2\theta)M_1M_2 = h\kappa M_1M_2$$

as introduced on p. 32 dominates the spin-spin coupling. This equation defines the effective coupling, coefficient κ which reaches 40 kHz if r is small. For the spectrum of the first nucleus, the selection rules are $\Delta M_1 = 1$, $\Delta M_2 = 0$ and the problem is entirely analogous to the spin-spin coupling in ^{13}C–H (Figure 3.6 and p. 33) with κ replacing J. Two lines are observed at $\nu \pm \kappa/2$ whose frequency separation is κ.

As with spin-spin coupling, the formulae require modification if the nuclei are both of the same species, but again two peaks are obtained. The frequency separation is $3\kappa/2$ not κ when $\mu_1 = \mu_2$ and $I_1 = I_2$.

A simple case is that of the two protons in each water of crystallization of gypsum, $CaSO_4 . 2H_2O$. Only the coupling between hydrogens of the same water are important, but for most orientations of a single crystal θ will be different for the two water molecules and so four resonance peaks are obtained. Figure 3.11 shows the experimental pattern for three

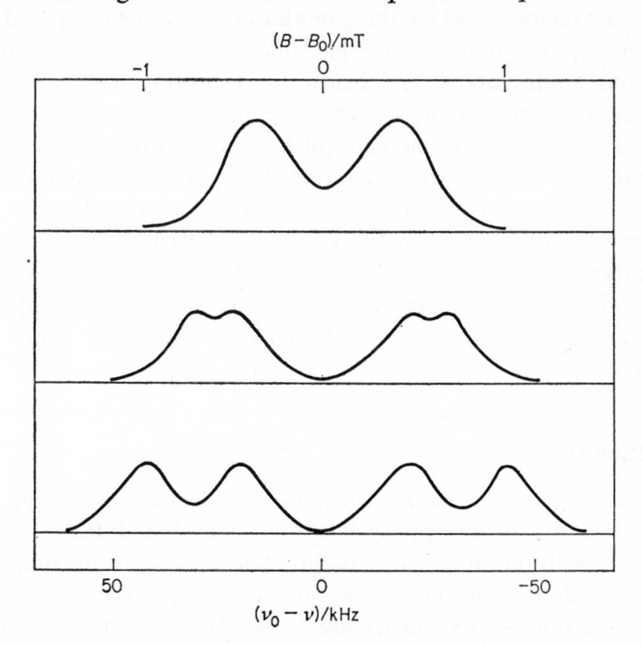

3.11 Nuclear magnetic resonance absorption of hydrogen nuclei in gypsum, $CaSO_4$, $2H_2O$, single crystal for three different crystal orientations.

different orientations. The maximum separation will be $(3\mu_0\mu_H^2/\pi r^3 h)$ when $\theta=0$ since $I=1/2$ for hydrogen. The hydrogen-hydrogen distance r may be accurately obtained by this means and an analysis of the orientational dependence of the lines gives information about the orientation of the water molecules in the crystal.

Analogous treatments are possible for the three hydrogens of an H_3O^+ ion and other species with more than two hydrogens. If the material is polycrystalline each value of θ occurs for some individual crystals and considerable line broadening results. However, the probability of any value of θ varies as $\sin\theta$ which means that $\theta=\pi/2$ is the most probable. The separation for this value predominates and two broad maxima occur with approximately this separation. Using the correct statistical treatment accurate values of r may be extracted from the observed curves. This method of determining hydrogen-hydrogen distances in solids is of considerable importance since these are not accurately obtained by X-ray diffraction methods.

Intermediate in character between rigid solids and liquids are those solids in which rotation occurs either intramolecularly—as with many methyl groups—or by molecular rotation about one or more axes in the crystal. Nuclear resonance line shapes show marked transitions in the temperature regions at which these motions became fast compared to the width, in frequency units, of the rigid solid pattern. Detailed comparison of the line widths above and below the transition may indicate the precise nature of these motions. Thus the line width (more precisely the square root of the second moment) of benzene falls from 310 to 130 μT in the range 90 – 120 K, which indicates the onset of rotation about the sixfold molecular axis. Complete rotation about all axes does not occur until the melting-point when the line width falls to less than 10^{-7} T. Many other solids and especially polymers show some rotational motion in the solid state which can be detected by this means and the temperature of the anomaly compared with specific heat anomalies, glass–rubber transitions, etc.

Relaxation times

It has been assumed above that the population of the nuclear spin orientation states are those appropriate to the field and temperature in accordance with the Boltzmann distribution for systems in equilibrium. However, such equilibrium may be slow to establish itself since most of the forces acting on the nuclei are weak. The rate of attainment of equilibrium may be expressed by a relaxation time, T, such that the departure from macroscopic equilibrium magnetization is proportional to exp $(-t/T)$.

Nuclear processes are governed by two relaxation times. The first, denoted T_1, is called the *spin-lattice* relaxation time or the *longitudinal* relaxation time and refers to the magnetization parallel to the magnetic field. If processes occur in times shorter than T_1 equilibrium may not be established. Thus there may be a delay in the appearance of a nuclear resonance signal of this magnitude after the magnetic field is switched on. Likewise the enhanced magnetization of a large magnetic field may be obtained in a weak field, such as the earth's magnetic field, by a rapid reduction from the larger field. Also if large radiofrequency measuring power is used, saturation of the signal may occur. The oscillator energy absorbed by the spin system increases the number of spins in the excited state. If the number of quanta absorbed per second is more than the number of spins which return to the ground state by virtue of both the induced emission and the relaxation mechanisms, the states tend to become equally populated so that the absorption which is proportional to the population difference is diminished.

By means of this power saturation, temperature disequilibrium is established between the spin degrees of freedom and the thermal bath. Such temperature disequilibrium may persist for times of the order of T_1. It is possible by certain dodges, such as sufficiently rapid field reversal, to obtain more nuclei in the excited state than in the ground state. Normal nuclear resonance experiments on systems in such disequilibrium give emission signals.

The second relaxation time, T_2, is often called the *spin-spin* or the *transverse* relaxation time. It governs the distribution of nuclear magnetization in the plane perpendicular to the magnetic field. T_2 is approximately half the reciprocal of the line width in rad s^{-1} and may be measured in this way. More accurate measurements of T_1 and T_2 can be made with special 'spin-echo' nuclear resonance equipment, which uses pulsed radiofrequency fields. It is not always possible to describe the relaxation of real systems in terms of only two exponential constants and quite sophisticated treatments may be necessary, especially when non-equivalent nuclei or nuclei with spins greater than 1/2 are present.

An essential difference between the relaxation steps governed by T_1 and T_2, is that T_1 processes involve a transfer of energy from the spin system to the surrounding medium or lattice. In contrast T_2 processes are adiabatic and must be less restricted so that $T_2 < T_1$. For many mobile systems $T_2 \sim T_1$, whereas for solids with broad lines $T_2 \ll T_1$.

For solids T_2 may be as short as 10^{-8} s, whereas for mobile liquids it may reach a few seconds. T_1 may be as short as 10^{-4} s in the special cases of quadrupolar nuclei and of solutions of paramagnetic salts, but it is more commonly many seconds in pure liquids and may even be many hours in solids below 4 K.

Not all molecular processes affecting T_1 have yet been fully studied, but the more important involve molecular rotations and translations which give rise to fluctuating magnetic fields. These are radiofrequency fields and, like those used in the resonance experiments, are capable of inducing spin transitions. Being random in nature they lead to a statistical relaxation of the bulk nuclear magnetization. These fluctuating fields are more effective if the rates of molecular motions correspond to the correct frequency for transitions with the particular applied magnetic field. T_1 is thus field dependent. For fields with resonance frequencies of the order of 40 MHz, the efficient rate of molecular motion is encountered in very viscous liquids. Motions in mobile liquids are too rapid and in solids too slow to provide efficient relaxation. Curves of T_1 plotted against viscosity or temperature usually show a minimum of the order of 10^{-3} s.

Nuclear quadrupole resonance and quadrupole coupling

4

Definitions and energy expressions

Effects arising from the coupling of nuclear magnetic moments with magnetic fields were discussed in chapter 3. Nuclei with spin $I \geqslant 1$ also possess electric quadrupole moments, which are linked to the spin axis, and give rise to energy terms when they are in electric field gradients, especially those derived from valency electrons in the same molecule as the nucleus.

For nuclei, electric dipole moments are zero and the chief electrical term, apart from the charge itself, is the electric quadrupole moment. This may be thought of as describing the non-spherical shape of such nuclei. The spin axis is necessarily an axis of cylindrical symmetry, but the nucleus may be elongated along this axis, like a cigar, in which case the quadrupole moment is positive. Conversely some nuclei are flattened at the poles, like a discus, when the moment is negative.

The equation

$$Q = \int \rho r^2 (3 \cos^2 \theta - 1) \, d\tau$$

may be taken as a definition of Q, the quadrupole moment. ρ is the charge density per unit volume, r the distance of the volume element, $d\tau$ from the origin and θ the angle between the radius vector and the spin quantization axis; the integral is to be evaluated for the state for which $M_I = I$. ρ is normally expressed in units of the charge on the electron so that the product eQ appears in most practical expressions. Q has dimensions of a length squared and since nuclear radii are $\sim 10^{-14}$ m, Q are $\sim 10^{-28}$ m^2. For some values see Table 3.1 on p. 23.

The nuclear quadrupole moment interacts with the gradients of the electric field, E, in which it is situated. These gradients are minus the second derivatives of the electric potential, V. These quantities are commonly denoted by q with appropriate subscripts to indicate the directions. The z direction, where appropriate, is taken as that for the maximum field gradient and the absence of subscripts to q implies

$$q \equiv q_{zz} = -\frac{\partial E_z}{\partial z} = \frac{\partial^2 V}{\partial z^2}$$

Some other authors use eq for the quantity designated q in this chapter. The symbol q implies minus the field gradient at the nucleus in question: in its computation the contribution from the charge on the nucleus itself must be omitted.

For the solid state, in the absence of magnetic fields, the quadrupole coupling energy has the form

$$E = eq_{zz}Q[3M_I^2 - I(I+1)]\,[4I(2I-1)]^{-1}.$$

This energy is independent of the sign of M_I so that if $M_I \neq 0$ a twofold degeneracy remains. This may be removed by a magnetic field. Figure 4.1

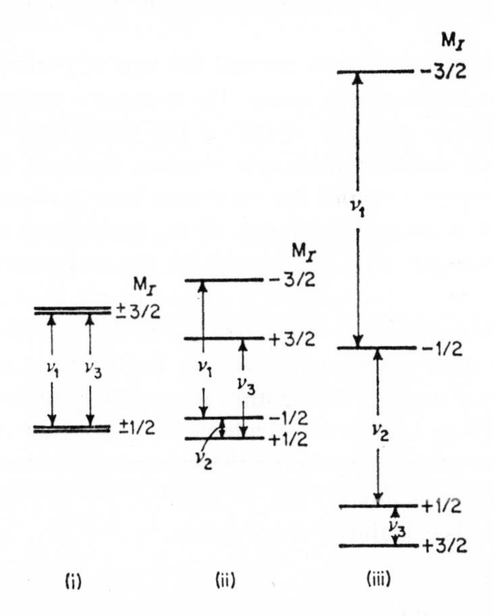

4.1 The energy level diagram for a spin 3/2 system with cylindrical symmetry and a magnetic field parallel with the cylinder axis. If eqQ is the quadrupole coupling and ν_N the nuclear resonance frequency in the absence of quadrupole effects, the transition frequencies are $\nu_1 = |h^{-1}eqQ + \nu_N|$; $\nu_2 = |\nu_N|$; $\nu_3 = |h^{-1}eqQ - \nu_N|$. The separate diagrams correspond to the cases (i) $\nu_N = 0$ and $h^{-1}eqQ > 0$, (ii) $6\nu_N = h^{-1}eqQ$. (iii) $3\nu_N = 4h^{-1}eqQ$.

exemplifies the changes caused by a field in the z direction; other cases are similar though they differ in numerical detail.

For direct observations of transitions in a radiofrequency magnetic field perpendicular to the quantization axis, $\Delta M_I = \pm 1$. This is precisely the selection rule for the nuclear resonance spectrum and is derived from the same interaction, namely that of the magnetic dipole to the radiofrequency magnetic field. The magnetic moment governs the transition moment and the intensity is independent of the quadrupole moment, except as this enters the Boltzmann distribution at thermal equilibrium. For $I=1$ there is one transition frequency, namely that referring to the degenerate transitions $M_I=0$ to ± 1. Likewise for $I=3/2$ there is the one observable transition between $M_I=\pm 1/2$ and $\pm 3/2$ levels. The allowed transition between $M_I=+1/2$ and $-1/2$ is at zero frequency. For $I=5/2$ the transitions between $M_I=\pm 5/2$ and $\pm 3/2$ and between $M_I=\pm 3/2$ and $\pm 1/2$ levels have different frequencies. The frequencies depend both on the nucleus through Q and on the molecular environment through q and each of these factors is capable of wide variation between different cases. Almost all transitions lie below 2 GHz and those below 1 MHz are difficult to observe because of unfavourable population differences.

When molecular rotations are present the axis of preferred quantization is changing its direction in space. The nucleus is trying to maintain its rotational axis in space by virtue of the gyroscopic effect of any spinning top, and simultaneously the nucleus is trying to align itself along the changing direction of the maximum field gradient. The consequences of this is a strong broadening of the quadrupole energy levels, which may be thought of as consequent on the uncertainty of nuclear orientation or on the rapid relaxation caused by the fluctuating electric field gradient when referred to space fixed axes. The net result is that no spectrum can be observed in liquid systems. For gases, at low pressures, the total rotational angular momentum is accurately quantized and the quadrupole coupling energy is calculable by averaging over the well defined rotational motion. Providing the hyperfine structure to the pure rotational spectrum (chapter 7) can be resolved and identified, accurate values of eqQ can be obtained.

Measurements on solids

The basic type of radiofrequency bridge circuit described for nuclear resonance is essentially sufficient for the detection of pure quadrupole resonances. However, this is inconvenient since a search must be made over a wide frequency range and circuits with one tuning condenser are to be preferred. Super-regenerative oscillators and other special electronic circuits have been developed, with favourable signal-to-noise ratios.

Since no magnetic field homogeneity is involved quite large samples, several grams in weight, can be used. Even so detection is quite difficult and it may be necessary to work at liquid helium temperature to reduce relaxation times and improve the population differences between the states. Once absorption has been detected the frequency of the oscillator can be accurately measured against frequency standards.

Interpretation of the spectra

There are several features which are common in pure quadrupole spectroscopy and are illustrated by the spectrum of solid tin tetraiodide, SnI_4. The iodine spectrum consists of four transitions giving two strong lines at 203·63 and 407·22 MHz and two more of about one-third the strength at 204·59 and 409·19 MHz. Iodine has only one natural isotope, ^{127}I, of spin 5/2, and the standard formulae given above indicate the energy levels and transition frequencies of Table 4.1. This shows that one

Table 4.1 *Quadrupole coupling energies and transition frequencies for a nucleus of spin 5/2*

ENERGY LEVELS		TRANSITIONS	
$M_I = \pm 5/2$	$(1/4)eqQ$	$5/2 \leftrightarrow 3/2$	$(3/10)h^{-1}eqQ$
$M_I = \pm 3/2$	$(-1/20)eqQ$	$3/2 \leftrightarrow 1/2$	$(3/20)h^{-1}eqQ$
$M_I = \pm 1/2$	$(-1/5)eqQ$		

of the two allowed transitions for each nucleus should be just double the other in frequency. The weaker lines at 204·59 and 409·19 MHz show this relationship accurately. The other lines show approximately this feature and must be assigned to other iodine transitions. In the crystal lattice, the SnI_4 are not perfect regular tetrahedra, but one iodine of each molecule lies on a three-fold symmetry axis and the remaining three iodine atoms are equivalent to each other and related by this axis. This symmetry is analogous to that in CH_3D. The weaker lines are attributed to the unique or axial iodine and the stronger lines to the other three.

Although $407{\cdot}22 \simeq 2 \times 203{\cdot}63$, the difference is greater than the experimental errors. The breakdown of the simple formula, whereby one frequency should be exactly twice the other, is ascribed to lack of cylindrical symmetry; then corrections must be made to the theory. The Sn–I bond along the threefold axis must have effective cylindrical symmetry, but for the others this is not strictly required by the symmetry. This asymmetry of the field gradient is described in terms of an asymmetry parameter, η, defined by

$$\eta = (\partial^2 V/\partial x^2 - \partial^2 V/\partial y^2)/(\partial^2 V/\partial z^2) \equiv (q_{xx} - q_{yy})/q_{zz}.$$

For SnI_4, $|\eta| = 0{\cdot}009$. Much larger values are known in other compounds, especially when the atom forms part of a double bond system as in iodobenzene. The signs of eqQ and η are not determined by these experiments although they are normally obtained from the microwave gas rotational spectra.

For the three equivalent iodine atoms $|h^{-1}eqQ| = 1357{\cdot}7$ MHz and for the axial iodine $|h^{-1}eqQ| = 1364{\cdot}0$ MHz. It is apparent that crystal forces are sufficient to change q by about 0·05 per cent as all four iodines would be fully equivalent in the vapour state. Local variations of crystal forces due to impurities, crystal strain, radiation damage, etc., give rise to line broadening through the variability of such crystal field effects: this may be so severe as to make the signal too small to observe.

The influence of crystal forces also manifests itself as a temperature coefficient, since the forces differ as the crystal expands or contracts. For SnI_4 the temperature coefficient is $-1{\cdot}00 \times 10^{-4}$ K^{-1} so that $h^{-1}eqQ$ at 77 K has fallen by 28·8 MHz from its value at 300 K. This is a large shift compared to the accuracy with which the frequency can be measured and to the line width of about 10 kHz.

Relaxation times

When quadrupole coupling is present relaxation times tend to be shorter than those observed for nuclei with spin 1/2. This relaxation may broaden hyperfine lines related to nuclei with quadrupole moments and even appear to remove the coupling to such nuclei in both nuclear magnetic resonance (chapter 3) and electron magnetic resonance (chapter 5) experiments.

When $I > 1/2$ there are at least 3 energy states and a complete description of the relaxation behaviour can be quite complicated. The new feature, which is added to those discussed in chapter 3, is the coupling of the quadrupole moment to fluctuating electric field gradients; in particular this allows transitions for which $\Delta M_I = \pm 2$. In principle such transitions could be observed in a spectroscopic measurement, but in practice it is not possible to design a condenser suitable for containing the sample and for producing large enough radiofrequency electric field gradients. However, on the microscopic scale the translation and rotation of individual molecules or ions give large, randomly fluctuating field gradients which can cause rapid relaxation of quadrupolar nuclei. As with the analogous magnetic fields in nuclear magnetic resonance, the most effective motions are those which occur near the transition frequency. Induced motions at this frequency can be generated by the application of ultrasonic waves and the extra relaxation in the presence of such waves has been demonstrated.

Interpretation of couplings

Providing the transitions can be identified, it is straightforward to derive the coupling energy eqQ from either solid or gas phase measurements. If Q is unknown it can be determined if q can be estimated or measured by the use of an isotope whose quadrupole moment is established. The ratio of the quadrupole moments of a pair of isotopes can be obtained with considerable accuracy. Such ratios, and also approximate quadrupole moments, play a valuable part in comparisons between experiments and the theory of nuclear structure.

More frequently Q is known and the measured value of q, the field gradient, provides information about the electronic structure of the molecule. Consider first the coupling in a free atom. For an electron in an s orbital there is spherical symmetry to the electron distribution and

$$\partial^2 V/\partial x^2 = \partial^2 V/\partial y^2 = \partial^2 V/\partial z^2 = 0,$$

since by Laplace's equation

$$\partial^2 V/\partial x^2 + \partial^2 V/\partial y^2 + \partial^2 V/\partial z^2 = 0.$$

Therefore s electrons provide no quadrupole coupling energy. The same argument applies to any closed shells of electrons since these too have spherical symmetry. Although the electron density for an electron in a p orbital vanishes at the nucleus, the field gradient does not. With a p_z orbital there is cylindrical symmetry about the z direction and

$$\partial^2 V/\partial z^2 = -2\partial^2 V/\partial x^2 = -2\partial^2 V/\partial y^2 \neq 0.$$

These relationships arise from the symmetry and the application of Laplace's equation. An electron in a d orbital is even less likely to be near the nucleus, since its wave function varies as r^2 for small r in contrast to the p orbital wave functions which vary as r. f orbital functions vary as r^3. Consequently q, which refers to the gradient at the nucleus, is very small for d or f electrons and can often be neglected compared to the values which arise from the p orbital distributions. In many cases the coupling due to exactly one p electron can be obtained from atomic beam experiments. In others approximate values can be obtained from a trial wave function. For a p_z electron $q = \langle e/5\pi\varepsilon_0 r^3\rangle_{\text{average}}$. If Q is accurately known $\langle r^{-3}\rangle_{\text{average}}$ can be obtained from the coupling. For two electrons both in the p_z orbital the gradient would be double; if two perpendicular orbitals are each occupied, the x, y, and z gradients must be separately added.

This theory for atoms may be transferred to molecules. The gradient from electrons centred at other nuclei and from these nuclei themselves is small compared to the effect of p electrons centred at the nucleus whose coupling is observed. Inner shells are spherically symmetrical and the molecular coupling is therefore almost entirely controlled by the occupancy of the p orbitals of the valence shell. The electrons shared in covalent bonds as well as the lone pair electrons must be included. U_p is often written for the fraction of unbalanced p electrons and is obtained from the measurements of the equation

$$U_p = -\text{observed coupling}/(\text{coupling for one p electron}).$$

Table 4.2 gives the observed eqQ for some chlorine derivatives and the value of U_p derived therefrom.

Table 4.2 *Some quadrupole coupling energies for* ^{35}Cl *and derived values of Up*

	$(h^{-1}eqQ/\text{MHz})$	*Up*
Atomic Cl	−109·6	1
KCl	± 0·04	4×10^{-4}
TlCl	− 15·795	0·14
CH_3Cl	− 74·740	0·68
CD_3Cl	− 74·41	0·68
ICl	− 82·5	0·75
BrCl	−103·6	0·94
FCl	−145·99	1·33

For gaseous KCl there is virtually no coupling, which is consistent with a spherically symmetrical Cl^- ion. It is interesting that the coupling should be so small, since the positive K^+ ion distorts the polarizable Cl^- ion, as shown by the reduction of the dipole moment to 82 per cent of the expected value for an ion pair. This shows that polarization effects occur predominantly at large distances from the nucleus and that the distribution close to the nucleus is less affected. For TlCl there is definite indication of a lack of p electrons near the chlorine which implies some degree of covalent bonding. This bonding is larger for CH_3Cl, ICl and BrCl, and for the latter the bonding appears to be close to a pure covalent bond. For FCl even more electron density has been transferred to the fluorine atom, which is consistent with the large electronegativity of fluorine.

Such qualitative explanations of the couplings are reasonably satisfactory and several attempts have been made to account more exactly for the couplings in terms of electronic structure. Unfortunately the quadrupole coupling, while it provides evidence for the presence or absence of p electrons, gives no indication of which orbitals the electrons occupy instead. For a detailed discussion it is necessary to know the ionic character of the bond, the overlap of the bond forming orbitals, the degree of s and d character in the bond and in the lone pair electrons, etc. Evidence on these points may be taken from dipole moments, bond energies via electronegativity scales, Hammett sigma values, analogous molecules and a variety of semi-empirical equations. Although some of these treatments give plausible descriptions, the arguments are not always unequivocal and each author is inclined to treat the data somewhat differently and no treatment appears clearly superior to the others.

Electron resonance 5

Principles

Many of the principles of nuclear resonance apply also in electron resonance. Like many nuclei, an electron has a spin S of 1/2 and an associated magnetic moment. Consequently in a magnetic field the two spin states, corresponding to the resolved spin, M_S, $= +1/2$ and $-1/2$, have different energies. Transitions may be caused by a high-frequency

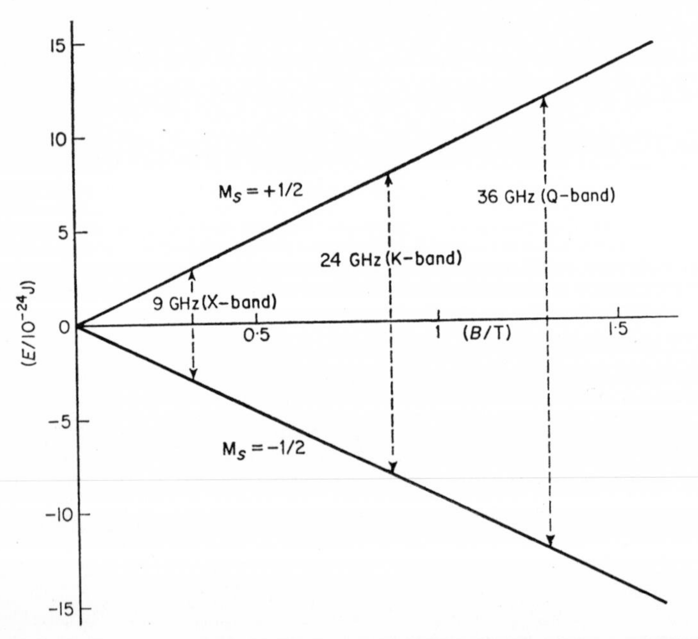

5.1 The energy and transition frequencies of an electron with g = 2·00 as a function of magnetic field. X-, K-, and Q-band are common names for the separate microwave ranges including the frequencies indicated.

magnetic field perpendicular to the steady field. The electron has a magnetic moment some 700 times that of the hydrogen nucleus and the transition frequencies are correspondingly higher. In a magnetic induction of 1 T, the frequency is 28 GHz which lies in the microwave range. The resonance condition is usually put in the form

$$h\nu = g\mu_B B,$$

where μ_B is the Bohr magneton $= e\hbar/2m_e = 9{\cdot}3$ m² A. g is the proportionality factor, which is sometimes called the spectroscopic splitting factor. For a free electron it is 2·0023 but this value is slightly modified for electrons in molecules. Figure 5.1, which should be compared with Figure 3.1, p. 25, indicates the energy relationship.

The most important difference between nuclear and electron resonance derives from the fact that in molecules electrons regularly occur in pairs with opposed spins as required by the Pauli exclusion principle. If one

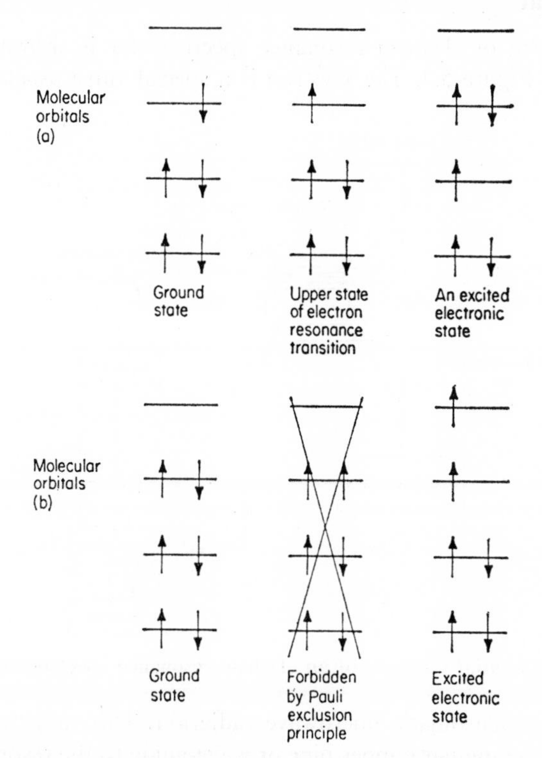

5.2 Electron shell filling for (a) an odd number of electrons, e.g. 5 and (b) an even number, e.g. 6. With non-degenerate orbitals there is an inevitable spin degeneracy in (a) which leads to electron resonance while in (b) such degeneracy is necessarily absent.

electron of such a pair were to have its spin direction reversed it could no longer remain in the same space orbital and would necessarily be raised to an empty electronic orbital of higher energy. This would require energy of many kJ mol^{-1} and this could only be extracted from visible or ultra-violet radiation. It is only when an unpaired electron is present that a transition may be induced by microwave radiation. Figure 5.2 shows the filling of molecular orbitals for five and for six electrons in a molecule and demonstrates this argument. Such unpaired electrons are present in many atoms, in organic radicals and a few known inorganic radicals and in many compounds and salts of transition metals, rare earth metals and trans-uranic elements. Signals may be obtained from small amounts of these materials in the presence of a large excess of spin paired molecules.

Experimental

A simple form of electron resonance spectrometer is shown diagrammatically in Figure 5.3. The klystron is a special valve oscillator which

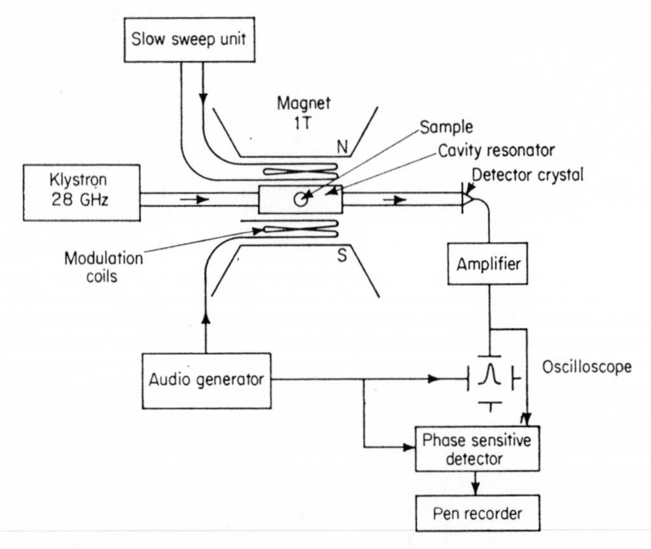

5.3 Essential elements of an electron resonance spectrometer.

produces monochromatic microwave radiation. This radiation may be led down a rectangular copper pipe or wave-guide to the resonant cavity in the magnetic field. The sample to be examined is placed in this cavity whose purpose is to increase the microwave magnetic field at the sample. The radiation which is transmitted by the cavity can be modified by the resonance absorption in the sample. This may be liquid or solid and normally occupies about 0·2 cm^3. The transmitted radiation is detected in a rectifier crystal and the signal appropriately amplified. Two forms of display are common. If the magnetic field is varied sinusoidally, at perhaps 50 Hz, by an amount which is larger than the absorption line width then the output shows a dip whenever the resonance condition is reached. This may be displayed on an oscilloscope with a related time base. Though convenient, this system has the disadvantage of an inherently poor sensitivity because of the large bandwidth required. To avoid this objection, presentation of the derivative of the absorption on a pen recorder is often preferred. For this a much smaller amplitude of field modulation is applied. This is less than the line width and the signal is proportional to the difference of absorption between the extremes of the modulation. See Figure 5.4. This is strictly proportional to the slope of

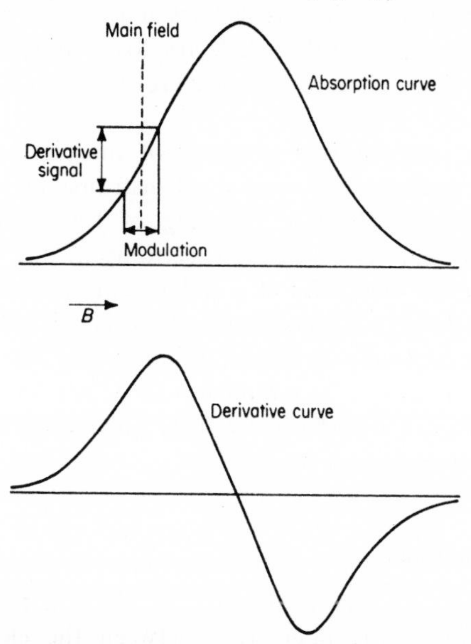

5.4 An electron resonance absorption curve and its first derivative which is obtained by field modulation and phase sensitive detection of the signal at the modulation frequency.

the chord, which is almost equal to the tangent at the mid-point for small modulations. The field of the magnet is slowly varied to cause this mid-point to traverse the absorption in a few minutes. With a phase detector at the modulation frequency the net output at the pen is the derivative of the whole absorption curve.

Results are often presented in the derivative form in diagrams in scientific literature. With practice they are readily interpreted. Every negative slope in the derivative indicates a peak or shoulder in the absorption spectrum. Each crossing of the axis of the derivative with a negative slope, as at the centre of Figure 5.4, represents a true maximum and a crossing with positive slope represents a minimum of the absorption.

Additional complexities in more sophisticated spectrometers are introduced to improve the frequency stability of the klystron and the signal-to-noise ratio of the system. For the latter reason it is common to use a much higher modulation frequency, often 100 kHz, or to use superheterodyne detection.

The large magnetic moment of the electron and the high frequencies required lead to better sensitivities than in nuclear resonance. Concentrations of 10^{-8} mol m^{-3} are not unusual and as little as 10^{-12} mol (10^{12} spins) can be detected in narrow lines at 4 K. Spin paired molecules may have an adverse effect if their dielectric loss is high, since this affects the cavity properties. Solutions in water are especially unfavourable for this reason.

The transition moment is a property of the electron and not the radical in which it is situated, and if all instrumental conditions were accurately determined absolute counting of free radicals would be possible. This is very difficult and spectrometers are calibrated with reference to weighed quantities of a stable radical such as $\alpha\alpha$-diphenyl β-picryl hydrazyl. Even then accuracy to ± 20 per cent is not easily achieved, although relative amounts in similar samples may be measured more accurately. The counting of free radicals can be important in estimating radiation damage, in following free radical polymerizations, in studying enzyme and other reactions, etc.

Hyperfine interactions

Some of the most interesting features of electron resonance of free radicals arise out of the interaction between the electron magnetic moment and nuclear magnetic moments. The spectroscopic theory resembles that of spin-spin coupling in nuclear resonance. An energy contribution haM_SM_I must be added to the interactions with the external field, and with the usual selection rules, $\Delta M_S = 1$, $\Delta M_I = 0$, a line appears at a frequency aM_I with respect to the central transition frequency. If $I = 1/2$, $M_I = +1/2$ or $-1/2$ and the spectrum consists of two lines at $\pm a/2$ with respect to the position expected were this hyperfine coupling absent. a may be expressed in MHz and is strictly the splitting in a spectrum taken with fixed field and variable frequency. If the spectrum is run at fixed frequency and variable field, a corresponds to the splitting in induction and is often quoted in such units. At $g \sim 2{\cdot}00$, 1 Tesla $\equiv$ 28 GHz. With more than one magnetic nucleus present the terms may be added independently.

The coupling constant, a, may be split into a direction dependent component,

$$h^{-1}\mu_0 g \mu_B \mu_I (4\pi I)^{-1} \langle (3\cos^2\theta - 1) r^{-3} \rangle_{\text{average}}$$

and a direction independent term. As discussed (p. 31) with the corresponding nuclear resonance term, the time average value of $(3\cos^2\theta - 1)$ in a liquid system is zero. This feature is not affected by the averaging over the electron wave function which is also required.

However, unlike the nuclear-nuclear interaction analogue, it is possible for r, the electron-nucleus distance, to be zero. Thus although the average of $(3\cos^2\theta - 1)$ is zero for a spherical electron distribution or for a liquid, the whole average remains finite because of the factor r^{-3} in the region $r \rightarrow 0$. The correct value in this limit is $(8\pi/3)\Psi^2(0)$ where $\Psi(0)$ is the value of the normalized wave function of the unpaired electron at the nucleus in question. This expression was first obtained by Fermi and the term is called the Fermi term or the contact term. p, d and f orbitals centred on the nucleus have zero values of $\Psi(0)$ but the use of s orbitals by the unpaired electron gives rise to large contributions.

For a free hydrogen atom the hyperfine coupling is 1420 MHz. If there is a lesser hydrogen coupling, x MHz, in a free radical one may say that the unpaired electron occupies a molecular orbital in which the hydrogen 1s orbital has a coefficient $(x/1420)^{1/2}$. Rather more crudely one may say the unpaired electron occupies the 1s orbital $x/14{\cdot}2$ per cent of the time or sometimes it is said that the spin population of the orbital is $x/1420$. Although not apparent from a one electron description it is possible for x to be negative if other electrons are taken into account. This implies that the predominant spin population of the s orbital is due to electrons of resolved electron spin opposite to that of the unpaired electron.

A specially common case of negative spin population occurs in simple planar systems of the form $X_2C{-}H$ which include most aromatic radicals such as $C_6H_6^-$. The π electron on the C atom, or aromatic ring system, polarizes the spin distribution of the pair of electrons forming the C–H bond in the sense which requires the electron of reversed spin to be near the hydrogen. Thus the wave function of the C–H fragment is a fractionally unequal mixture of those corresponding to the basic forms I and II of Figure 5.5. I is favoured by virtue of a more favourable exchange energy between the electrons of like spin on the carbon atom.

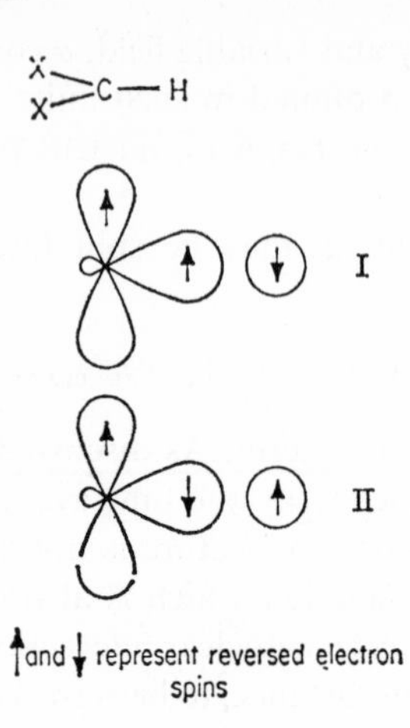

5.5 Origin of the negative hyperfine coupling in > C–H fragments. See text.

As an example of hyperfine coupling Figure 5.6 shows the spectrum of the semiquinone negative ion showing five peaks separated by 6·6 MHz

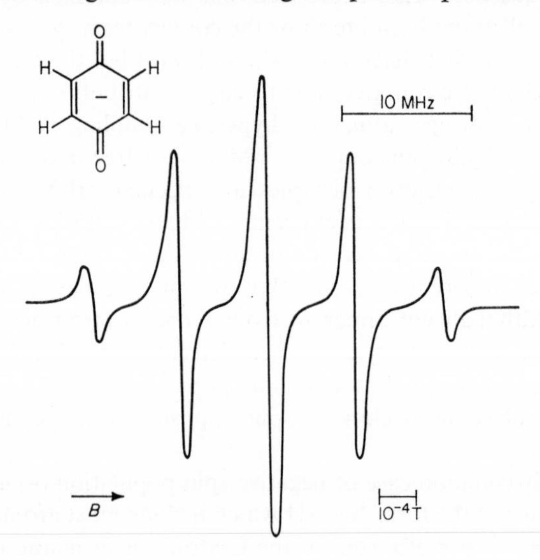

5.6 First derivative of the absorption curve of the electron spin resonance spectrum of the semiquinone radical anion.

with intensity ratios 1 : 4 : 6 : 4 : 1. Figure 5.7 indicates the energy levels and transitions involved. Direct coupling of the nucleus with the magnet has been neglected. The degeneracies indicated, which account for the

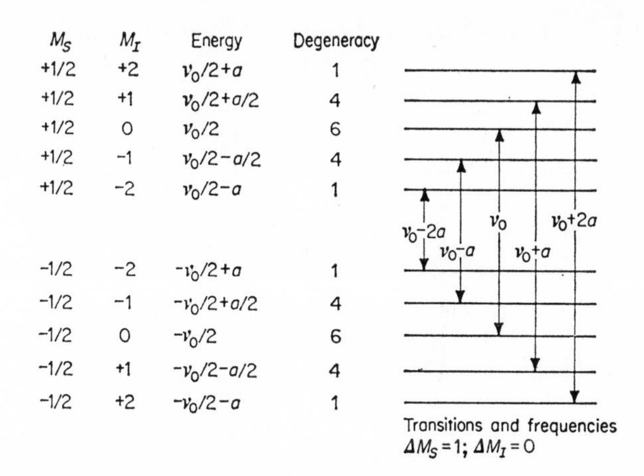

5.7 Energy levels and allowed transitions for the semiquinone radical anion in a magnetic field. $\nu_0 = h^{-1}g\mu_B B$

intensity ratios, are related to the number of alternative ways in which the corresponding values of M_I may be reached. Thus $M_I=1$ may be reached in four ways according to which hydrogen has resolved spin $-1/2$; the other three hydrogens must each have resolved spin $+1/2$ so that the net resolved nuclear spin, M_I, is $+1$.

Table 5.1 gives values of the hyperfine coupling constants for a number of radicals and shows the order of magnitude to be expected. In more complicated species isotopic substitution may be required to distinguish which hydrogen atom is coupling. In aromatic systems there is some justification for writing $a_H = Q\rho_C$, where ρ_C is the spin population of the p orbital of the carbon orbital adjacent to the hydrogen atom whose hyperfine coupling is a_H. Q is a constant which is roughly independent of the radical and equal to -63 MHz. Spin populations obtained in this way agree approximately with theoretical predictions, but it is doubtful if Q is sufficiently independent of charge, bond lengths, etc., for precise, reliable values of ρ_C to be obtained.

Although in solids the direction dependent term is not averaged to zero, it is often small compared with the isotropic or contact term. In such cases the solid lines are broader but the spectrum resembles those discussed for liquids and can be interpreted in terms of an average coupling constant. If it is possible to obtain oriented radicals as impurities in a molecular host crystal much sharper electron resonance lines are obtained, whose exact spacing varies with the orientation of the crystal with the magnetic field. Such orientation can be obtained by crystallizing radicals along with a host of the same general size and shape and also,

Table 5.1 *A selection of isotropic hyperfine coupling constants, (a/MHz)*

^{1}H IN		^{14}N IN	
H	+1420	NH_3^+	+ 54·6
CH_3	− 65	ND_3^+	+ 52·7
NH_3^+	− 72	NH_2	+ 29
HCO	+ 384	NO_2	+ 153
	(+398, +382, +370)		
HPO_2^-	+ 230	^{19}F in	
	(+238, +227, +224)	CH_3	+ 405
$C_6H_6^-$	− 10·5		
$C_{10}H_8^-$ (α position)	− 14·0		
(β position)	− 5·0	^{31}P in	
$HC(COOD)_2$	− 60	$PO_3^=$	+1665
	(−91, −61, −29)		(+1967, +1514, +1514)
$(CH_3)_2CCOOD$	+ 65	HPO_2^-	+1385
			(+1698, +1228, +1228)
^{13}C in			
CH_3	+ 115	^{55}Mn in	
CO_2^-	+ 468	Mn^{++}	− 270
	(+546, +436, +422)		
$H^{13}C(COOH)_2$	+ 93		
	(+213, +42, +23)		

(Where orientation effects have been studied the principal values of the coupling tensor are given in brackets.)

very conveniently, by damaging the molecules in a pure crystal by high energy X-rays, γ-rays, electrons, etc. Figure 5.8 shows the spectra for three orientations of irradiated glycollic acid in which the radical HOCH(COOH) becomes trapped. The four lines correspond to coupling to two hydrogens, the larger coupling being due to the hydrogen attached directly to the free radical centre and the other being due to the hydroxyl hydrogen.

In a more exact treatment the coupling information is expressed as a tensor. With strong anisotropy the selection rule $\Delta M_I = 0$ may not apply and in this case the hyperfine pattern becomes a function of the size of the magnetic field.

g-factors

g, in the resonance equation $h\nu = g\mu_B B$, represents the proportionality factor between the transition frequency and the magnetic field. The theoretical value of g for a free electron is 2·0023 and this has been confirmed by experiment. However, when the electron is bound in a radical,

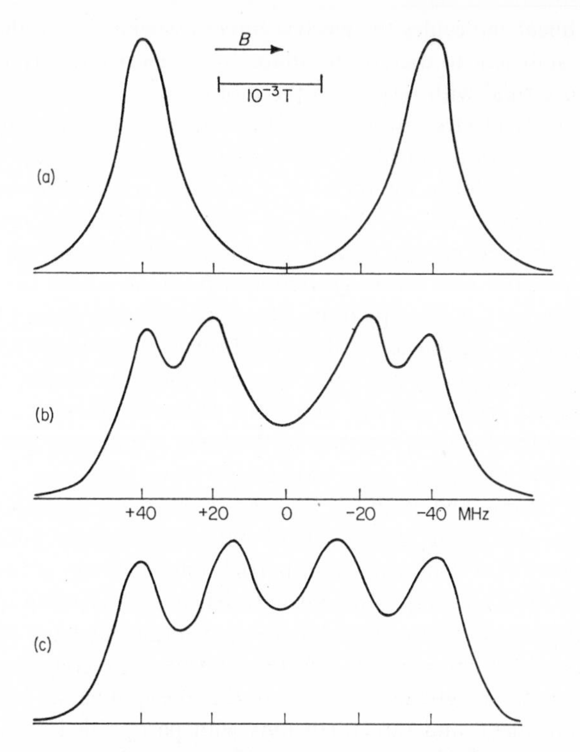

5.8 Electron resonance absorption of the radical HOCHCOOH trapped in a single crystal of irradiated glycollic acid, $HOCH_2COOH$. Three different crystal orientations relative to the magnetic field are shown. The approximate splittings due to H–C are (a) 80, (b) 60, (c) 58 and to H–O are (a) 0, (b) 10 and (c) 15 MHz respectively.

g departs from this value and the resonance is displaced in a manner analogous to a chemical shift in nuclear resonance. For organic radicals a typical *g*-factor would be 2·0043, which corresponds to a shift of the centre of the spectrum of 9 MHz or 3×10^{-4} T at 9 GHz. This is less than the overall width of most spectra with hyperfine couplings, so that if two or more chemical species of radical with different *g*-factors should be present, serious overlapping occurs. *g* is strictly a tensor and its effective value may vary with radical orientation.

The feature which gives rise to variations of *g* is the action of the electron magnetic moment which can induce orbital magnetic moments which in turn interact with the magnetic field. When the electron is reversed in the transition the induced orbital moments also change sign and the extra energy required must be extracted from the microwave quantum.

In non-linear molecules the electric forces associated with the positive nuclei are sufficient to dictate the shape of the electronic orbitals which are therefore rigid with respect to the molecular framework. This situation is in contrast to free atoms where there are no internal features which fix the directions of the orbitals. For three degenerate p atomic orbitals it is convenient to refer to the forms p_{+1}, p_0 and p_{-1} with resolved angular momentum along z of $+1$, 0 and -1 respectively. In discussing compounds the real forms p_x, p_y and p_z, or hybrids thereof, are the most suitable basis. p_{+1} may be recognized as a p orbital whose nature varies from p_x to p_y to $-p_x$ to $-p_y$ to p_x, etc., cyclically with time. The orbital motion is said to be quenched if the real forms p_x, etc., are required in contrast to the imaginary $p_{+1} \equiv p_x \pm ip_y$. Non-linear molecules or radicals have quenched orbital angular momenta. However, there is a spin-own orbit interaction for each electron of the form $\zeta \boldsymbol{l}.\boldsymbol{s}$: here $\boldsymbol{l}$ and $\boldsymbol{s}$ are the electron orbital and spin angular momentum vector operators. $\zeta = 29\ cm^{-1}$ for carbon atoms and is much higher for elements of larger nuclear charge. The spin-orbit term is small but not negligible compared to electrostatic forces and causes a partial unquenching of the orbital momentum. The magnetic moment induced by each electron is opposed to its own intrinsic moment, but since some electrons have $M_S = +1/2$ and others $-1/2$ there are both positive and negative contributions and as a result g may be greater or less than the free spin value, 2·0023.

Transition metal and rare earth ions with partly filled d and f shells and with only weak electrostatic forces have especially large departures from $g = 2{\cdot}0023$ and a very wide search for a resonance must be made.

Triplet and higher multiplet spin states

In the discussion so far it has been implicitly assumed that the electron had two spin states, $M_S = \pm 1/2$ whose energy difference was a linear function of the magnetic field. This is only correct for doublet states, that is those with only one unpaired electron as in typical organic free radicals. With two or more unpaired electrons the situation is slightly different. In most cases the total spin $S = s_1 + s_2 + s_3 \ldots$ is quantized and the state of lowest energy is that for which the individual electron spins are parallel; then $S = n/2$ where n is the number of unpaired electrons. In the absence of a magnetic field there is a $(2S+1) = (n+1)$-fold degeneracy associated with the possible directions of S. The names singlet, doublet, triplet, quartet, quintet, etc., refer to this degeneracy and commonly arise with 0,1,2,3,4 . . . unpaired electrons respectively.

Though the degeneracy is not removed by spherically symmetric electric fields, which occur in atoms, more asymmetric electric fields are able to remove the degeneracy so that there are energy separations even in zero magnetic fields. The size of these zero field splittings vary rather widely with the type of molecule or metal ion, but are typically of the same order as the splitting due to a strong magnetic field. Figure 5.9

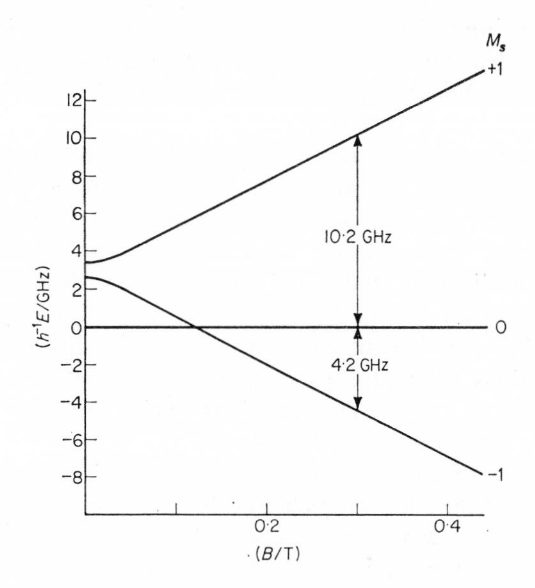

5.9 Energy versus magnetic field strength parallel to the z direction for naphthalene in its triplet state. The transitions for $\Delta M_S = 1$ at 0·3 T are indicated.

shows a typical energy diagram and indicates how unequal transition frequencies arise for any given magnetic field. These frequencies vary with the orientation of the magnetic field so that the transitions are best observed in single crystals where the molecules are aligned. M_S is not a good[1] quantum number although it can be used to label the levels at intermediate fields. The $\Delta M_S = 1$ transitions are the strongest but $\Delta M_S = 2$ transitions can also be observed. Resonances from triplet states in electronically excited naphthalene and related molecules have been observed and shown to have a decay rate equal to that of the phosphorescence.

The energy as a function of spin direction is usually expressed in the

[1] *'Good' quantum numbers* are the values of quantum numbers related to observables which are independent of time for the undisturbed system. They are regularly exact integers or half-integers. In mathematical terminology the observable commutes with the Hamiltonian. Under other circumstances an analogous quantum number description may be convenient for labelling states even though the quantum numbers no longer provide an exact description.

form

$$hDS_z^2 + hE(S_x^2 - S_y^2)$$

or

$$-h(XS_x^2 + YS_y^2 + ZS_z^2), \text{ (with } X + Y + Z = 0)$$

where S_z, etc., are to be read as spin operators. Figure 5.9 is drawn for the triplet excited state of naphthalene for which $D = +3018$ MHz and $E = -414$ MHz. For the triplet ground state of Ni^{++} in $NiSO_4.7H_2O$, $D = -108$ GHz and $E = -45$ GHz.

It might be thought that transition metal ions were spherical so that the zero field splitting was absent, as in gaseous atoms. However, the surrounding molecules or ions in the crystal exert sufficiently strong influences to give large splittings. These neighbours are sometimes called ligands or coordinating groups and quantitative predictions about the details can be made on the ligand field theory. The details are quite involved and for the ions of larger atomic number there may be spin-orbit interactions which are also important. The net effect of ligand fields, etc., may be to move the transitions outside the microwave range and make the spectra awkward to observe.

For ions with an odd number of electrons there is *Kramer's theorem* which states that the effect of electric fields can never remove the degeneracy completely. Always the states occur in pairs with reversed sign of resolved angular momentum. This degeneracy can be removed by a magnetic field and so transitions in the microwave range can be observed in magnetic fields of a few tesla.

An extreme case of unpaired electrons can occur in some solid metals where essentially a large proportion of the outer electrons are aligned. This is the basic arrangement of ferromagnetic and permanently magnetic materials, although the details are confused by the effects of domain structure, grain boundaries, dislocations, etc. The magnetic resonance in such materials is extremely broad. Resonance may also be obtained from the conduction electrons in other metals in the colloidal state. The corresponding resonance in bulk metals is not easily observed because the microwave fields do not penetrate electrical conductors to more than a very small skin depth.

If the magnetic centres are somewhat further apart, as in organic crystals of 100 per cent free radicals, the forces which tend to align the electrons parallel to each other are weaker and are overcome by thermal motions. In such solids, of which $\alpha\alpha$-diphenyl β-picryl hydrazyl is the common example, the electrons are fairly independent and strong electron resonance signals can be observed. In such systems spin exchange can occur between adjacent radicals; that is if the spins are anti-parallel they can both reverse their spin directions by a mutual process. The coupling of the electrons to individual nuclei is then averaged out since the unbalanced electron spin visits a large number of different radicals during the time required to establish a spectroscopic transition of, say, 5 MHz line width. Associated with this loss of hyperfine structure is a narrowing of the resonance line. This exchange narrowing and loss of hyperfine structure may also occur in strong solutions.

Relaxation times, etc.

Two relaxation times, T_1 the longitudinal time and T_2 the transverse time, are required to describe the electron resonance process and these are entirely analogous to the corresponding quantities in nuclear resonance. T_1 may be from 10^{-7} s to a few minutes according to the system and the temperature; the long times are found at liquid helium temperatures. T_2 usually lies at the shorter end of this range and is essentially the reciprocal of the line width in radians per second.

One contribution to T_2 in some cases is an electron exchange reaction. For benzophenone negative radical ion this reaction can be written

$$(C_6H_5)_2CO^-(a) + (C_6H_5)_2CO(b) \rightarrow (C_6H_5)_2CO(a) + (C_6H_5)_2CO^-(b)$$

where individual molecular frameworks have been artificially labelled (a) and (b). In the electron jump represented by this reaction the hyperfine coupling is disturbed since (a) and (b) have, in general, different detailed arrangements of resolved nuclear moments of their hydrogen atoms. Consequently for sufficiently frequent exchanges the hyperfine coupling is no longer observed in the spectrum. The electron jump rate may be deduced from a quantitative study of the line widths.

6 Mössbauer spectroscopy

Although nuclear spectroscopy is outside the scope of this book, there are some aspects of the Mössbauer effect which relate closely to the type of molecular information which is gained by magnetic resonance spectroscopy. No attempt is made in this chapter to cover aspects of the Mössbauer effect which are not connected with molecular energy levels.

Principles of the Mössbauer effect

A number of excited radioactive nuclei emit γ-rays and thereby reach the ground state of the nucleus with the same mass number and same charge, and so of necessity of the same chemical element. If such γ-rays are allowed to interact with a similar ground state nucleus, the γ-ray may be reabsorbed in a resonance capture process. This possibility has been recognized for some time, but was expected to be an inefficient process. This was because, although the greater part of the nuclear transition energy appeared in the γ-ray, a small but significant fraction may remain as nuclear recoil energy. Such energy is associated with the nuclear velocity and more particularly with the momentum required to balance, vectorially, the momentum carried by the γ-ray. Similarly the momentum balance of the reabsorption process requires the nucleus to possess excess kinetic energy after the reabsorption. Consequently the emitted and absorbed γ-ray should differ by twice the kinetic energy involved and this would not be readily supplied by thermal energies. But the expectation was disproved in 1958 by Mössbauer who showed that resonance capture of the γ-ray could occur with moderate efficiency if the source and the absorber were both contained in solid lattices. This is possible because in a proportion of the absorption and emission processes the momentum is taken up by the entire crystal rather than by the individual nucleus. Since the kinetic energy, $mu^2/2 = (\text{momentum})^2/2m$

is inversely proportional to the mass for a given momentum, this energy is essentially zero when the mass is raised by a factor of perhaps 10^{18} from that of the nucleus to that of a small crystal. A fuller understanding of why such 'recoilless' transitions should occur, and of the quantitative aspects, requires a discussion of the phonon nature of the lattice motions and is not essential for an understanding of the applications.

The energies of the γ-rays from such recoilless emissions are very precise. The spread of energies approaches a theoretical limit set by the Heisenberg uncertainty principle. This spread is thus related to the lifetime of the excited state, a quantity which is more often called the radioactive decay constant. The 14·4 keV γ-emitting isomer of ^{57}Fe is a very important nucleus for Mössbauer experiments and this has a decay constant $\tau = 2 \times 10^{-7}$ s and so the full width of the energy spread at half the maximum intensity is $(\pi\tau)^{-1} = 1{\cdot}6$ MHz in frequency units. This is of the order of magnitude of, or less than, most of the energies encountered in magnetic resonance. If then the emitting and absorbing nuclei were in different nuclear orientations, or perhaps in different chemical compounds, the exact energy match would not be achieved and reabsorption would not occur. If some means were available to an experimenter for changing the energy of the γ-ray by a few MHz, the resonance equality might be restored. If the small energy adjustments were themselves changed systematically with time a full absorption spectrum could be investigated. Such a process is available through the Doppler effect. If the emitter or source and the absorber are in motion with a relative velocity u, the frequency or energy of the γ-ray appears in the reference frame in which the absorber is at rest to be changed by $\Delta\nu$ or ΔE where

$$\Delta E/h = \Delta\nu = \nu u/c = Eu/ch.$$

If $E = 14{\cdot}4$ keV, the γ-ray energy, and $u = 1$ cm s^{-1}, then $\Delta\nu = 116$ MHz. This change is rather too large for most situations so that velocities up to 1 cm s^{-1} are normally sufficient and these are easily produced. The raw data of a Mössbauer spectrum is thus the relative number of γ-rays per second passing the absorbing sample as a function of the relative velocity of the source and sample.

One feature which reduces the usefulness of the Mössbauer effect for molecular spectroscopy is that not all elements have suitable nuclear transitions. Nuclei which do have suitable excited states include ^{57}Fe, ^{61}Ni, ^{67}Zn, ^{119}Sn, ^{129}I, ^{161}Dy, ^{169}Tm, ^{197}Au.

Experimental

Many details of the experimental arrangement vary with the precise experiment envisaged and possibly with the individual components available. The source is usually a thin foil, perhaps a tenth of a millimetre thick, containing the radioactive nuclei. The absorber is a film of similar thickness covering the window of the counter. This may detect the γ-rays by the ionization they produce, or the γ-rays may be transformed to light pulses in scintillation counters. Many sources emit other high-energy radiation than the γ-ray of interest and a proportional counter with an energy selector arranged to count only γ-rays of the desired energy is then essential. A long count of several hours may be required to bring out the spectrum above the noise due to the statistical uncertainty in the arrival time of the γ-rays. Such noise is proportional to the square root of the number of counts and if only 10^4 counts were made at each velocity there would be a precision in the absorption of 1 per cent of the count. This is scarcely accurate enough if the absorption itself is only 3 per cent or so at the peak of the resonance. Strong sources with favourable geometrical arrangements may give counts of several hundred per second so that over two minutes would be required for each velocity and a spectrum with a resolution of one-hundredth of the velocity range studied would require over three hours.

Sometimes the relative motion, perhaps a mm s^{-1} or so, is obtained by mounting the source or absorber on an engineering lathe bed and driving from the lead screw. However, it may be difficult to keep the standard counting rate constant over large distances of travel or over long times, particularly if the precursor of the γ-emitter has a fairly short half-life. There is always likely to be such a precursor since the half-life of the γ-emitter itself must be 10^{-5} s or less for efficient resonant capture. As in other forms of spectroscopy the relationship between the probability of spontaneous emission and of absorption are related by the expressions given in chapter 2. Because of difficulties in constancy, it is usual to scan through the velocity range several times a second. The source is then attached to a moving microphone coil whose motion is controlled electrically to provide the required velocity drive, ideally a saw tooth variation of velocity with time. The counts at each velocity are fed, over equal time intervals, into separate channels of a multichannel analyser as directed by a control signal appropriately derived from the electrical drive to the microphone. At the end of the counting period of a few hours the total count in each channel represents one point on the velocity spectrum. Figure 6.1 shows this scheme in the form of a block diagram.

There is a slight temperature dependence of the frequency which is related to a second order Doppler shift of the order of u^2E/c^2 which can become appreciable if u is a thermal velocity and, for the most accurate

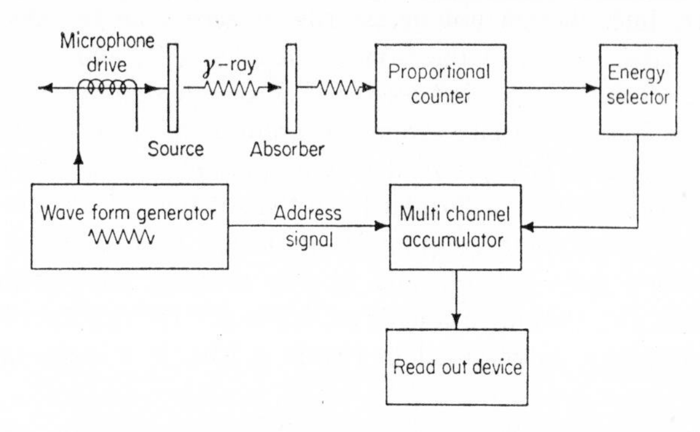

6.1 Block diagram of apparatus for Mössbauer effect.

shift measurements, a thermostatic control of source and absorber temperature may be required. Also the recoilless fraction of the γ-rays increases with a decrease of temperature. This fraction is 0·70 for ^{57}Fe at 300 K which is exceptionally favourable, but for ^{197}Au the value is less than 10^{-4} and only reaches 0·20 at 4 K. If the absorbing isotope is not present in 100 per cent abundance enrichment in the desired isotope gives a better ratio of absorption to background scattering. For ^{129}I where the isotope does not occur naturally such enrichment is essential.

Interpretation of the spectra

The possible forms of the observed spectra depend on the spins of the two nuclear states involved and on the selection rule for γ-ray emission as well as on the symmetry and magnetic state of the environment. The case of ^{57}Fe, 14·4 keV emission will show the possibilities. The excited nucleus has a spin of 3/2 and the ground nucleus a spin of 1/2. The γ-ray transition is essentially a magnetic dipole transition; this knowledge is required to determine the selection rules and relative intensities.

The symmetry and nature of the surroundings of the nucleus determine the degeneracy of the energy levels by virtue of the degeneracy of the nuclear orientation levels. If the surroundings are diamagnetic and spherically symmetrical, or of a high symmetry such as cubic or octahedral, the nuclear orientation degeneracy is not removed and only one emission or absorption frequency is possible. Rather complex spectra arise if both source and absorber have many transition frequencies and it is common practice to choose a source with only a single line. ^{57}Co, the precursor of ^{57}Fe, is usually incorporated in stainless steel or platinum foil to provide a suitable single line source. If the absorber also has only one transition frequency the Mössbauer spectrum consists of a single line, though not necessarily at zero relative velocity. The shift of position is called the isomer chemical shift and its molecular origin is discussed in the next section. The energy diagram corresponds to Figure 6.2(i). If the site symmetry around the absorber nucleus is reduced, an electric field gradient may be present and the energy of the $M_I = \pm 3/2$ orientations will differ from that of the $\pm 1/2$ states because of electrical quadrupole coupling as indicated in Figure 6.2(ii). Two transitions are now possible and in this instance the selection rules indicate that they should be equally probable and the two lines should be of equal intensity as observed in Figure 6.2(ii). If a magnetic field is

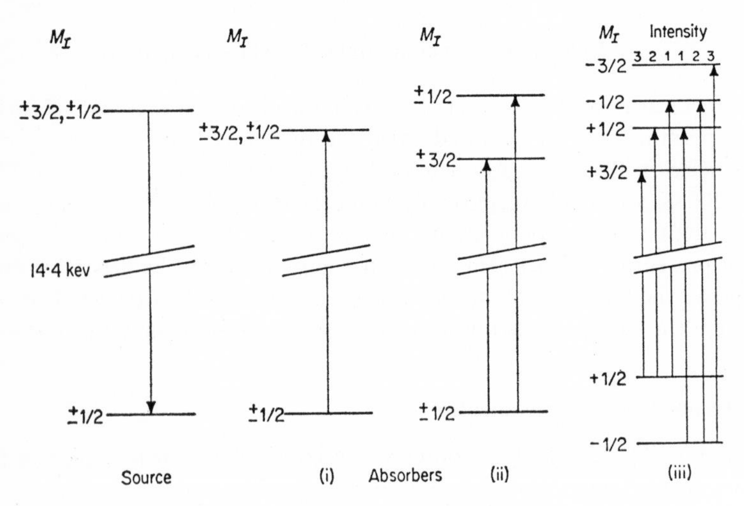

6.2. Energy diagrams for nuclear transitions in ⁵⁷Fe. See text.

also present the remaining degeneracy is lost and six transitions are possible as in Figure 6.2(iii), with relative intensities 3 : 2 : 1 : 1 : 2 : 3 as indicated. The $\Delta M_I = \pm 2$ transitions are almost forbidden in this case. The magnetic field to produce this splitting is the local effective field at the nucleus. Though this can be produced by an external magnet, it may also be the field internal to the domains of a ferromagnet. In this case the hyperfine coupling is very large and modified by the co-operative nature of the ferromagnetism. Though in principle it may be measured by nuclear magnetic resonance, the metallic nature of the specimen makes this difficult on bulk samples and thin samples modify the ferromagnetism, so that the Mössbauer effect, for which electrical conduction causes no difficulty, has advantages in this field. Isolated paramagnetic centres have splittings related to the eigenvalues of the hyperfine coupling Hamiltonian such as those discussed in chapter 5.

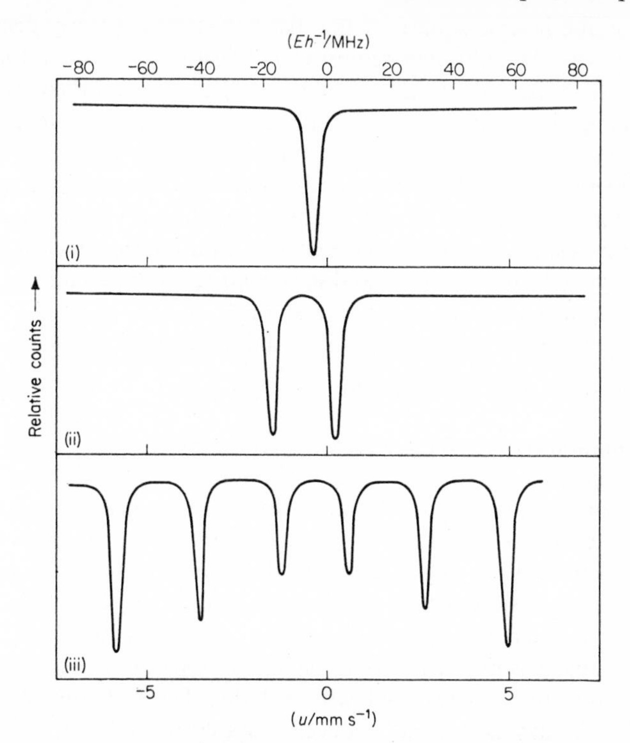

6.3 Some Mössbauer effect spectra. (i) $K_4Fe(CN)_6$, (ii) $Na_2Fe(CN)_5NO$, (iii) metallic iron absorbers, referred to a source of ^{57}Co in platinum.

Isomer shifts

The energies obtained from the spectra related to the quadrupole coupling can be interpreted as discussed in chapter 4 and the magnetic splittings follow the arguments of chapter 5. If as in Figure 6.3(iii) the splitting is observed in a ferromagnetic material the energy gaps of Figure 6.2(iii) are then $\hbar\gamma B_{\text{eff}}$, where B_{eff} is the effective magnetic induction. This must be the same in both states so that accurate ratios of $\gamma_{\text{excited}}/\gamma_{\text{ground}}$ can be obtained.

It is then the shift of the origin that is a novel molecular effect, which is only encountered elsewhere in high resolution atomic spectroscopy; in this latter field, unlike in the Mössbauer effect, the nuclei may be of different mass number and the phenomenon is better known as an isotope shift in atomic spectroscopy. The origin of these effects lies in the finite effective radius of the nucleus and the breakdown of the concept of a point charge at the centre. The Coulomb law of attraction varying as the inverse square of the distance is consequently disturbed and the precise Hamiltonian requires modification. The wave function of the electrons near the nucleus is likewise disturbed and the correction to the energy is a function of the effective nuclear radius, R, and the value of the square of the wave function at the origin, $\Psi^2(0)$. The difference in energy of one electron when the nucleus enlarges by δR is then proportional to $\Psi^2(0)\,(\delta R/R)$ and this must be summed over all electrons. This energy is small compared to the energy of the internal nuclear energy accompanying the change of δR and so is not directly apparent. But if two different electronic arrangements are studied for each nucleus then the difference or shift becomes observable and is proportional to:

$$[(\Sigma\Psi^2(0))_a - (\Sigma\Psi^2(0))_b](\delta R/R).$$

The summation is over all the electrons present. For the Mössbauer effect a might represent the absorber and b the source. The difference of the two γ-ray energies required is equal to this shift and is measured by the energy shift supplied by the Doppler effect to restore resonance. For atomic spectroscopy a might represent the ground state and b an electronically excited state, when the shift is equal to the difference in the two atomic frequencies for the two isomers or isotopes concerned. For a source of mixed isotopes it would equal the separation between their corresponding lines as resolved with a Fabry–Perot etalon or similar device. There may be a small reduced mass correction, but the major part of the isotopic shifts arises from the different nuclear radii.

For the atomic cases there may be good electronic wave functions based on self consistent field computer programmes and so the $(\delta R/R)$ may be estimated. For molecules and solids as used in the Mössbauer effect, the wave functions are less well known. Nor consequently are the values of $(\delta R/R)$ well known. Since $(\delta R/R)$ is a constant for a given nuclear transition, values of $\Sigma\Psi^2(0)$ can be compared on a relative scale. $\Psi^2(0)$ is only non-zero for s electrons, since p wave functions have a node at the nucleus. For most elements only the outer s electrons are disturbed on compound formation and so the isomer shifts indicate the distribution of these valence s electrons. The sign convention is that positive velocities represent source and absorber moving towards each other and so positive isomer shifts indicate more energy in the absorber transition than in the source transition. If δR is positive, then positive shifts indicate a greater density of electrons at the nucleus of the absorber than at the source nucleus.

Table 6.1 shows the isomer shifts of some tin compounds. The divalent stannous absorbers are seen to have consistently much more positive shifts than the stannic compounds. An identification of the tin valency

Table 6.1 *Isomer shifts*

	STANNOUS		STANNIC	
Oxide	SnO	+ 2	SnO_2	−48
Sulphate	$SnSO_4$	+27	$Sn(SO_4)_2$	−54
Fluoride	SnF_2	+16	SnF_4	−53
Chloride	$SnCl_2$	+21	$SnCl_4$	−43
Bromide	$SnBr_2$	+26	$SnBr_4$	−25
Iodide			SnI_4	−23

All in MHz relative to ^{119}Sn in white tin as source.

could thus be made from the Mössbauer effect of a compound where the valency was uncertain. For ^{119}Sn $(\delta R/R)$ is positive and so the two extra electrons in the stannous compounds should lead to more positive isomer shifts, as is observed.

For ^{57}Fe the best estimate of $(\delta R/R)$ is -18×10^{-4} so that the more negative shifts represent the greatest s character in ^{57}Fe absorbers. Nevertheless it is the lower valency ferrous compounds which are found experimentally to have the most positive shifts. Thus ferrous sulphate, $^{57}FeSO_4 . 7H_2O$, has a shift of $+10{\cdot}8$ MHz and the hydrated ferric sulphate, $^{57}Fe_2(SO_4)_3 . xH_2O$, has a shift of only $+1{\cdot}6$ MHz with respect to ^{57}Fe in platinum as source in each case. The lower value of $\Sigma\Psi^2(0)$, associated with the higher total number of electrons in this instance, is attributed to the effect of the extra electron which must enter the 3d shell and partially shield the valence s electrons from the nuclear charge. The d electrons do not themselves contribute to $\Sigma\Psi^2(0)$ because of their node at the origin.

7 Molecular rotation

Energies and simple spectra

In chapter 1 it was stated that molecules could be considered to possess energy associated with their rotational motion which was relatively independent of any other forms of energy they might possess. In the gas state this rotational energy is quantized. For linear molecules the form of the energy is especially simple being given by

$$E_{rot} = J(J+1)\hbar^2/2I.$$

In this expression I is the moment of inertia and J, a positive integer; J is a characteristic of the energy level in question and is called the rotational quantum number. $J(J+1)\hbar^2$ is the square of the rotational angular momentum. It is common to write B for $h/8\pi^2 I$ so that the energy expression reads

$$E_{rot}\, h^{-1} = J(J+1)B.$$

B is called the *rotational constant* and may be expressed in MHz or cm^{-1}. These energy levels form a diverging series as shown in Figure 7.1. The resolved angular momentum along a specific, of z, direction may be quantized and the further quantum number, M_J, is required fully to specify the rotational state. $M_J\hbar$ is the resolved angular momentum along z and M_J may have any integral value from $+J$, $+J-1$, $+J-2$, $\ldots 0 \ldots, -(J-1)$, $-J$. There are $2J+1$ possible values of M_J and the Jth energy level is therefore $(2J+1)$-fold degenerate. In the absence of electric and magnetic fields, the energy does not depend on M_J or the direction in space about which the molecule is rotating.

A general, non-linear, molecule possesses three moments of inertia I_A, I_B and I_C, with associated rotational constants A, B and C respectively. By convention $I_A \leqslant I_B \leqslant I_C$ so that $A \geqslant B \geqslant C$. There is an important class of molecules, symmetric top molecules, for which two of the moments of inertia are equal, as in CH_3Cl. If $I_A \neq I_B = I_C$

$$E_{rot}\, h^{-1} = J(J+1)B + K^2(A-B)$$

if $I_A = I_B \neq I_C$ then

$$E_{rot}\, h^{-1} = J(J+1)B + K^2(C-B).$$

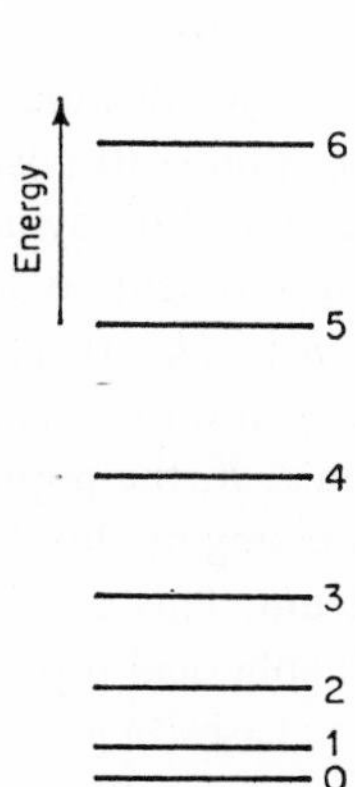

7.1 The rotational energy levels of a linear molecule.

K is the quantum number associated with rotation about the molecular symmetry axis. The internal angular momentum is $K\hbar$ about this axis. The energy, for a given J, increases or decreases with K^2 according as the two equal moments of inertia are greater or less than the third. K is restricted to integer values in the range $+J$, $+J-1$, $+J-2, \ldots, 0, \ldots, -(J-1)$, $-J$. Since the energy depends on K^2 the $\pm K$ levels are degenerate. This degeneracy is essentially associated with the clockwise or anti-clockwise rotations about the symmetry axis and is additional to the $(2J+1)$-fold degeneracy associated with the independence of the energy on M_J. These formulae also apply to spherical tops, for which $I_A = I_B = I_C$, and the energy reduces to

$$E_{rot} = hJ(J+1)B$$

which is the same as for linear molecules. However, there is here an extra $(2J+1)$-fold degeneracy, arising essentially from the $(2J+1)$ values of K, which has no counterpart in linear molecules.

Molecules for which $I_A \neq I_B \neq I_C$ are called asymmetric top molecules. The energy levels cannot be expressed in simple closed form and need to be computed numerically in each case or else interpolated from tables. The theory has been studied extensively and numerical tables are available for a wide range of parameters and the assignment of a large number of asymmetric top spectra has been successfully accomplished. Since the detailed theory is quite complicated and no new basic principle is involved it will not be given here. J remains a good quantum number.

Transitions in rotational energy are caused through the interaction of the electric field of the electromagnetic radiation with the molecular electric dipole moment, p. The selection rule in J is $\Delta J=0, \pm 1$ for transitions generally. For pure rotational transitions $\Delta J=0$ is only relevant for asymmetric tops, since in other cases the change of energy would be zero. $\Delta J=+1$ corresponds to absorption and $\Delta J=-1$ to the accompanying induced emission. For symmetric tops, the symmetry requires the molecular dipole moment to lie parallel to the symmetry axis and the selection rule in K is $\Delta K=0$. The transition moment is proportional to p and so the absorption intensity varies as p^2 with proportionality factors depending on J, K, the population of the states and line shape factors. Molecules possessing no dipole moment give zero intensity and no pure rotational spectrum. This applies to homonuclear diatomic molecules, as H_2, N_2, and all spherical top molecules, as methane, CH_4, as well as a number of other cases, e.g. ethylene, C_2H_4, and benzene, C_6H_6.

For linear or symmetric top molecules the frequencies of the $J \leftarrow J-1$ transition is readily evaluated since

$$h\nu = E_J - E_{J-1} = hB[J(J+1)-(J-1)J],$$

whence $\nu = 2BJ$.

The spectrum should consist of a set of evenly spaced lines with separation $2B$ and indeed essentially this is observed. Table 7.1 gives the frequencies of the known lines of OCS. Simple inspection shows that $2B$ is slightly over 12 GHz and that the J values of the left-hand column must be correct, although there are some missing transitions whose frequencies have not been measured. The frequencies are known to seven or eight significant figures and the expression $\nu = 2BJ$ does not hold to this accuracy and centrifugal stretching effects have to be included to obtain exact agreement. (See also problem 17.)

Introduction of numerical values shows that for small molecules and low J, the transitions occur near 3×10^{10} Hz $= 1$ cm^{-1} and the experi-

Table 7.1 *Observer rotational transitions of* $^{16}O^{12}C^{32}S$

$J \leftarrow J-1$	(ν/MHz)
$2 \leftarrow 1$	24 325·921
$3 \leftarrow 2$	36 488·82
$4 \leftarrow 3$	48 651·64
$5 \leftarrow 4$	60 814·08
$8 \leftarrow 7$	97 301·19
$10 \leftarrow 9$	121 624·63
$12 \leftarrow 11$	145 946·79
$14 \leftarrow 13$	170 267·49
$16 \leftarrow 15$	194 586·66
$18 \leftarrow 17$	218 903·41
$20 \leftarrow 19$	243 218·09
$22 \leftarrow 21$	267 529·56
$24 \leftarrow 23$	291 839·22

mental techniques required involve microwave spectrometers. For light molecules, especially those with only one atom other than hydrogen, the strong transitions lie in the far infra-red at 30–200 cm^{-1}. This region is experimentally more difficult and the resolution and frequency precision is inferior to that obtainable in the microwave range. However, the work of Czerny on HCl in 1925 was of historical importance as the results were conclusively in favour of the new quantum theory, which predicted energy levels proportional to $J(J+1)$ in contrast to an older form of the theory from which the energies would be proportional to J^2.

Microwave gas spectrometers

The accuracy of microwave measurements stems from the availability of klystron generators. These supply about 30 milliwatts of microwave radiation at a frequency which can be varied over a narrow range by electrical control and over a wider range by a mechanical setting. At a fixed setting their output is highly monochromatic. Their frequency can be accurately measured against the harmonics of stable quartz crystal controlled oscillators; these standard oscillators may be in the laboratory or may be those used to control the signals broadcast by certain radio stations for such scientific purposes. Klystrons are readily available from 3–50 GHz and weaker signals up to 250 GHz may be obtained with harmonic generators. There is thus a wide, useful frequency range although the lower frequency end is less favourable on grounds of sensitivity and the larger physical size of absorption cell required.

A simplified block diagram of an absorption spectrometer is indicated in Figure 7.2. The klystron generator delivers the radiation into a wave-guide—that is a hollow metal pipe. This literally guides the electro-magnetic radiation which is confined to the inside which contains the

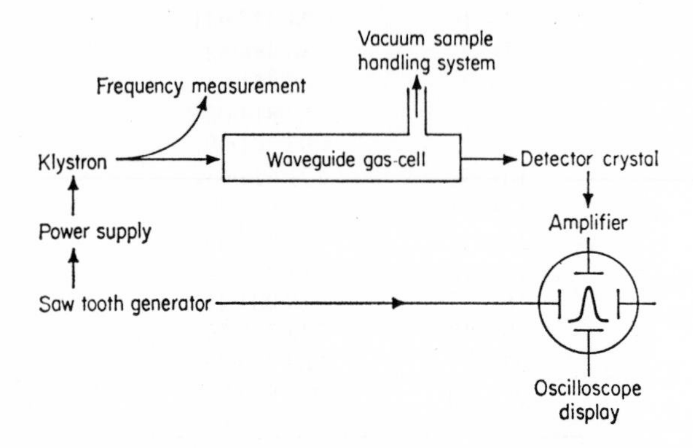

7.2 Simple gas microwave spectrometer.

sample gas. This cell may be up to 10 metres in length and has mica windows to contain the gas. This should be at a pressure $\sim$1 Pa for obtaining narrow lines and the whole cell may be cooled to improve the population of the lower energy states. After traversing the sample the radiation is detected in a crystal diode, the output amplified and fed to an oscilloscope. The exact klystron frequency is varied electrically by means of a sawtooth voltage applied to its reflector and this is also applied to the horizontal plates of the oscilloscope. The klystron is thus varied from its mean value, say 25 000 MHz, by a few MHz at a repetition rate of, say, 30 Hz. The klystron power is fairly independent of frequency and any sharp change at the detector is indicative of absorption by the sample and this can be observed on the oscilloscope. If none appears at the initial setting the centre frequency is varied by mechanical tuning and a neighbouring range examined.

Such a spectrograph is usually used for measurements of the highest accuracy as the absorption lines are narrow and fairly faithful in shape and relative intensities. The oscilloscope presentation means that the amplifier bandwidth cannot be narrowed to remove noise, so that the sensitivity is not exceptionally high. Also the search for new lines can be tedious unless their frequencies are known within narrow limits.

Some disadvantages of the simple absorption spectrometer are over-come in the *Stark modulation spectrometer* at the expense of some loss of resolution. The main items (see Figure 7.3) are similar except that the absorption cell contains a central electrode running the entire length of

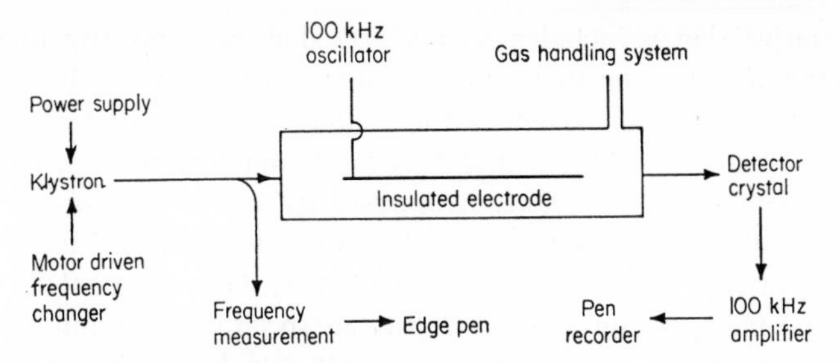

7.3 Simple Stark modulation gas microwave spectrometer.

the cell and insulated from it. To this is applied a high frequency (*c.* 100 kHz) square wave voltage, which is zero based. That is for half the cycle there is no voltage difference between the electrode and the guide and absorption of the microwaves by the gas occurs at the normal frequency. For the other half-cycle there is a large voltage which shifts and broadens the gas absorption (by virtue of the Stark splitting described below) so that the microwave power is no longer absorbed. If the frequency is set for gas absorption, a 100 kHz signal will appear at the detector crystal and may be amplified at this frequency which lies in a favourable operating range. In the absence of a gas absorption line no signal at 100 kHz reaches the detector. After amplification and detection the signal can be fed to a pen recorder. As the klystron frequency is driven by a mechanical motor a frequency range may be covered conveniently and the pen recorder left to indicate what absorption, if any, has been covered. Frequency markers can be arranged to actuate an edge pen on the recorder to show the numerical value of the absorption frequency. The damping circuit after the detector controls the effective bandwidth of the system which may be as low as 1 Hz; this is much smaller than could be used with an oscilloscope and enables a more favourable sensitivity to be reached.

More complex features of the spectrum

The theory of the opening section of this chapter suggests that the spectra should consist of a simple set of lines at frequencies given by $2BJ$, at least for linear molecules and symmetric tops. Many simplifications have been made in arriving at this result.

First, the intervals between the main lines are not given exactly by $2B$ because of the centrifugal forces in the rotating molecule which cause it to distort and so change its moment of inertia. The dominant correction term to the energy is of the form

$$-D_J hJ^2(J+1)^2.$$

The negative sign is included so that D_J shall be a positive quantity, since large J are associated with an increased moment of inertia and lower energy than that given by the simple formula. From a series of lines, D_J can be evaluated and thence B obtained more accurately. Since the centrifugal stretching is opposed by interatomic forces, D_J is a function of the force constants and its value may be used to obtain a relation between these constants, which may be additional to those obtainable from the infra-red and Raman spectra. For symmetric top molecules there are further distortions related to K and the small terms $-D_{JK}hK^2J(J+1)$ and $-D_K hK^4$ must be added to the energy. For transitions with $\Delta J=1$, $\Delta K=0$, the D_{JK} term gives a transition frequency dependent on K^2. The line is then split into a series of J closely spaced lines for the transition $J \leftarrow (J-1)$. If this K-splitting is clearly resolved it may be used to find J, if this is uncertain.

Another small term in the energy expression has the form

$$-h\Sigma_i\, \alpha_i(v_i+1/2)J(J+1).$$

This allows for the variation of moment of inertia with vibrational amplitude. In this expression the summation is over all the fundamental vibrations and v_i is the vibrational quantum number of the ith fundamental mode (see chapter 10). At room temperature or below almost all molecules have all $v_i=0$ and there is a simple correction to B of $-\Sigma_i\, \alpha_i/2$. The majority of α_i are positive, but α_i can be negative. The individual molecules which do not have all $v_i=0$ will have different effective B and will give rise to new absorption lines. Such lines can be distinguished using measurements at two temperatures, since they are weaker relative to the main lines at low temperatures, when the $v_i \neq 0$ states are less populated. If all the α_i can be determined by identifying lines for $v_i \neq 0$, the term $\Sigma_i\, \alpha_i/2$ is known and it is possible to determine B for the fictitious 'equilibrium' state in which even the zero point energy has been removed. This quantity is written B_e in contrast to B_0 for the ground vibrational state so that

$$B_e = B_0 - \Sigma_i\, \alpha_i/2.$$

Detailed expressions can also be determined for rotation-vibration interaction when other angular momenta are present and l-type doubling or Coriolis interaction is prominent. If the vibration is more properly described as a hindered rotation more complex interactions can occur, but in favourable cases the spectra can be unravelled and barrier heights and shapes obtained.

Another common cause of line multiplicity is the presence of isotopically substituted molecules. Since moments of inertia are functions of atomic masses, such molecules have a new set of moments of inertia and transition frequencies. The approximate positions of these lines can be estimated from approximate molecular structures and their intensities estimated from isotopic abundance ratios. If the isotopic lines are observed and identified, the new moments of inertia provide very useful data for determining molecular structure. Uncommon isotopes are, therefore, often introduced intentionally by chemical synthesis.

Yet another cause of multiplicity of lines arises from the interaction of the total rotation and nuclear orientations. The rotational angular momentum, J, and the nuclear angular momentum, I, must be added to give the total angular momentum which is governed by a new quantum number, F. The square of the total angular momentum is $F(F+1)\hbar^2$ and, as with other angular momenta, there is a $(2F+1)$-fold degeneracy in free space, the energy being independent of the resolved angular momentum $M_F\hbar$, where M_F takes the integer or half-integer values $+F$, $+F-1$, $+F-2, \ldots -(F-1)$, $-F$. F itself may take each of the values from $(J+I)$ to $|J-I|$ in integer steps. F is integer or half-integer according as I is integer or half-integer. If more than one nucleus has non-zero spin they must all be compounded in the formation of F. The selection rules have the form $\Delta F=0, \pm 1$, with the transitions for which $\Delta F=\Delta J$ the strongest.

For nuclei of spin 1/2 in the absence of external magnetic fields, the energy is almost independent of the relative directions of I and J. There is, however, a small interaction between the nuclear magnetic moment and the magnetic field generated by the rotating molecule and there is an energy term $hc\,\boldsymbol{I}\cdot\boldsymbol{J}$ in which c may be only a few kHz. The effects of this term can only be seen with the highest resolution.

Nuclei with a larger spin have a nuclear electric quadrupole moment and as seen in chapter 5 the corresponding coupling of the nucleus to the molecular framework may be as high as 1000 MHz. The coupling energy for a linear molecule with one such nucleus is given by

$$-eqQ[(3/4)C(C+1)-I(I+1)J(J+1)]\,[2I(2I+1)(2J-1)(2J+3)]^{-1},$$

where

$$C=F(F+1)-I(I+1)-J(J+1).$$

Since the energy is a function of F and a series of values of F are possible it is clear that quite a complicated spectrum, as Figure 7.4, may result. Extensive tables of both the energies and transition intensities have been compiled and it is comparatively straightforward, though tedious, to match the theoretical spectrum with the observed and so determine the quadrupole coupling constant, eqQ. The sign of this quantity as well as its magnitude is obtained.

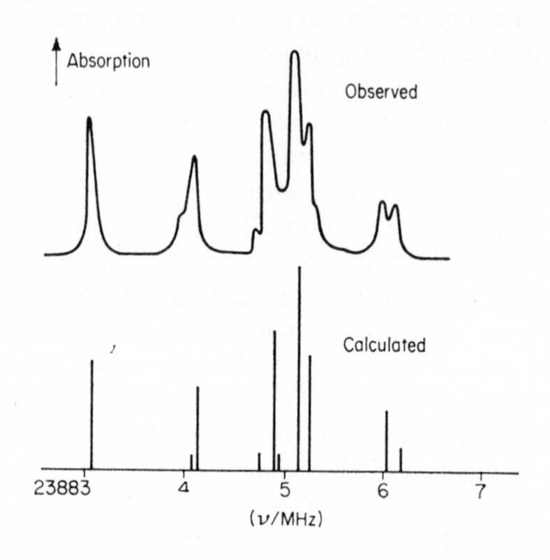

7.4 The high resolution gas microwave spectrum of $^{35}Cl^{12}C^{14}N$ near 23 885 MHz.

The spectrum of the linear molecule cyanogen bromide, BrCN exemplifies the type of complexity which may occur even in a simple molecule. With average resolution and sensitivity there are twenty-two lines between 24 465 and 24 885 MHz all of which are ascribed to the J 3←2 transitions; they are all at least one-hundredth of the intensity of the strongest line at 24 713·05 MHz. Eleven are ascribed to each of the isotopic species ^{79}BrCN and ^{81}BrCN. For each species there are identified five bromine quadrupole hyperfine lines for the ground vibrational state, two for molecules with one vibrational quantum in the carbon-bromine stretching mode and four for molecules with one quantum in the bending mode which has added complexity due to the vibrational angular momentum or *l*-type doubling. Under the highest resolution the stronger lines can be resolved into a further hyperfine pattern with separation of the order of 1 MHz due to the nitrogen quadrupole moment. Under higher sensitivity lines due to ^{13}C containing molecules would be detected.

Terms analogous to those in this section apply for asymmetric top molecules, cases of more than one nucleus with quadrupole coupling, etc.

Stark effect

Spectra in all frequency regions may be changed by the application of a strong electric field. These changes are termed *Stark effects* after their discoverer in the field of atomic spectroscopy. In rotational gas spectroscopy the electric field, E, interacts with the molecular dipole moment to give an energy term in linear molecules

$$E_{\text{Stark}} = p^2E^2[J(J+1) - M_J^2]\,[2BhJ(J+1)(2J-1)(2J+3)]^{-1}.$$

This is a second order Stark effect as the energy is proportional to the square of the electric field.

The selection rule is $\Delta M_J = 0$ for a static electric field parallel to the electric field of the microwave radiation and $\Delta M_J = \pm 1$ if the electric fields are perpendicular. The spectrum observed therefore depends on M_J as well as J. The number of transitions observed may be used to determine J and the separation to determine p. Line separations up to 20 MHz may be obtained in large fields and they can be measured to 20 kHz.

For symmetric tops the splitting is larger by a factor of the order of hB/pE as the energy is given by

$$E_{\text{Stark}} = -pEM_JK[J(J+1)]^{-1}.$$

This splitting is now first order, being linear in the field. The splitting removes the degeneracy in $\pm K$ states and the lines appear uniformly about the undisplaced centre except for the unsplit $K=0$ transitions. Although the energies and splitting are linear in the dipole moment, p, it is not possible to obtain the sign of p since the sign of K is not known for each individual transition.

Numerical values of the dipole moment can be obtained from these Stark effects with considerable accuracy. The greatest difficulty is to measure the electric field and this is usually calibrated from the Stark effect of a well-known dipole moment, such as that of carbon oxysulphide, OCS. The advantages of such microwave determinations of dipole moments may be summarized: (i) the accuracy is high; (ii) very small moments can be measured, e.g. propane, C_3H_8, for which $p = 3 \times 10^{-31}$ m s A; (iii) no density or pressure measurement is involved; (iv) exact temperature is unimportant and no temperature variation is required; (v) no corrections for atomic or electronic polarization are required; (vi) impure samples may be used, providing the required lines can be correctly identified; (vii) it is possible to determine the direction of the dipole moment with respect to the axes of inertia of asymmetric molecules; (viii) the measurement is normally made on molecules known to be in the ground vibrational state; (ix) it is possible to measure excited states and isotopic species and obtain the variation of dipole moment with vibrational motion. It may be possible to correlate this information with dipole moment derivatives from infra-red intensity measurements and in particular to determine the unknown signs required to interpret the intensity measurements.

Zeeman effect

Since Zeeman first observed the changes in atomic spectra in the presence of magnetic fields, magnetic field effects in spectroscopy are called *Zeeman effects.* A magnetic field interacts with magnetic moments, the largest of which are likely to arise from unpaired electrons or unquenched orbital angular momenta. Only a few molecules possess either of these properties in the ground state, although O_2, NO and ClO_2 and NO_2 do. These interactions can be as large as the rotational energies and quite complicated situations arise. Since these molecules are not typical, they will not be discussed except to comment that the interaction of the unpaired electrons of the triplet ground state of oxygen allows this molecule to have an extensive magnetically allowed microwave spectrum even though its electric dipole moment is zero.

In molecules where the electronic angular momentum is zero small terms arise from the molecular magnetic moments of the rotating states. Such a moment arises since the nuclei with their positive charges are at a different distance from the centre of gravity from the effective distance of their associated electrons. Rotating charges, like currents, produce equivalent magnetic moments and in the molecular case, those due to the nuclear motion and the electron motions do not cancel exactly although they are of nearly equal magnitudes and of opposite signs. The energy must be written

$$E_{\mathbf{Zeeman}} = -g\mu_{\mathrm{N}} M_J B,$$

where μ_{N} is the nuclear magneton, the usual unit for small magnetic moments, g is the molecular g-factor and $\boldsymbol{B}$ the magnetic induction. g is usually rather less than one and in strong fields the Zeeman energy may reach several MHz. The selection rule is $\Delta M_J = 0$ if the electric component of the microwave radiation is parallel to the magnetic field and $\Delta M_J = \pm 1$ if it is perpendicular, as with the Stark effect. The $\Delta M_J = +1$ and -1 transitions can be differentiated with circularly polarized microwave radiation and the sign of g determined. If nuclei with magnetic moments are present they too interact with the magnetic field with energies of this magnitude and more exact treatments must be used.

Inversion spectra

The first microwave spectrum to be observed, by Cleeton and Williams in 1934, was the inversion spectrum of ammonia. It lies at about 0·8 cm^{-1}, which equals 24 GHz, and was also the first spectrum to be examined by modern techniques in 1946 by Bleaney and by Townes.

This spectrum is concerned with the motion of the nitrogen atom between the centre of the three hydrogen atoms, as indicated in the inset pictures of NH_3 in Figure 7.5. The graph of potential energy plotted

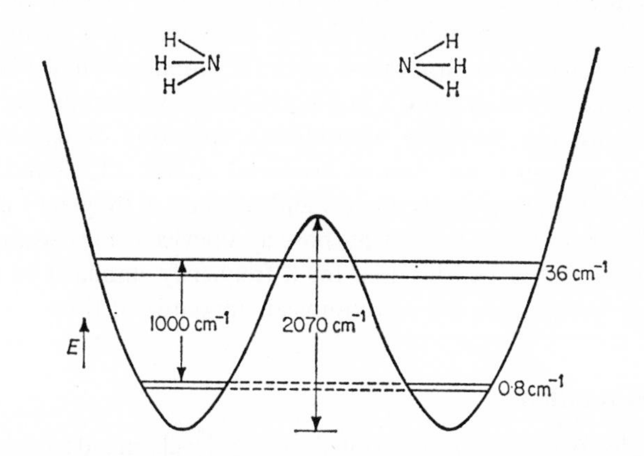

7.5 Energy diagram for the inversion coordinate of ammonia.

against the position of the nitrogen atom is of the double minimum type shown with a central barrier of 2070 cm^{-1} (25 kJ mol^{-1}). The curvature of each minimum gives the force constant for the simple motion, which is best described as a NH bending fundamental and lies near 1000 cm^{-1}. However, by virtue of the finite barrier height and the low mass of the hydrogen atoms there is an appreciable quantum mechanical tunnel effect which causes a splitting of the lower levels. The microwave transition is between the two components of the ground vibrational state. The transition moment is high so that the absorption is strong. Because of centrifugal effects the interval depends on J and K and there is a many line spectrum, which also shows the influence of the nitrogen quadrupole coupling.

The inversion frequency is strongly dependent on the reduced mass so that for ND_3 it lies near 0·05 cm^{-1} or 1600 MHz. For AsH_3 the calculated value is 1/2 cycle/year while for molecules such as CH_3Cl where bond rupture would be involved the frequency would be lower still. Under such circumstances the collision frequency exceeds the transition frequency and non-resonant absorption, analogous to rotation in the liquid state, is produced.

There is an experiment, which is interesting and, though difficult, is a practical proposition for the ammonia inversion spectrum. In appropriate, strong, inhomogeneous electric fields a gaseous beam of ammonia molecules behaves so that only molecules in the excited state of the inversion transition pass down the axis of the field electrodes. Consequently an absorption cell, preferably a microwave cavity, can be filled with molecules in the excited state. Absorption cannot occur, but induced emission is favoured and so it is possible to extract more microwave energy than is introduced. That is, the system acts as an amplifier or *maser* as such devices are called when the amplification is obtained by the stimulated emission from a non-equilibrium population distribution. With suitable gas pumping, continuous operation is feasible and a microwave oscillator can even be produced. Under such conditions the emission line is exceptionally narrow and constant in frequency. Although not quite as favourable for this purpose as a device using caesium atoms, the ammonia maser can be used for a frequency standard or clock for measuring variation in the astronomical standards of time.

Hindered rotation

The simple formulae for rotational energy levels are derived with the implicit assumption that the molecule is essentially rigid and that there is no free rotation of one end of the molecule with respect to the other. However, methyl groups are commonly free to rotate against the rest of the molecule and other examples of relatively free rotation are known such as that about the N–N bond in hydrazine, NH_2NH_2. Analysis of the microwave spectra, where such rotation does occur, is quite difficult, especially if one or both groups have dipole moment components which are perpendicular to the common axis. When such spectra can be analysed, accurate information may be obtained about the height and shape of the barriers opposing the rotation. For instance in methyl alcohol, CH_3OH, the height of the threefold cosine barrier is 4·47 kJ mol^{-1} (375 cm^{-1}) and it is established that the oxygen atom is displaced by 0·08 Å from the axis of the methyl group. The sixfold barrier in nitromethane, CH_3NO_2, is much lower at 25·1 J mol^{-1} (2·10 cm^{-1}) while the smaller zero point vibrational amplitudes of CD_3NO_2 lead to a reduction of barrier height to 21·6 J mol^{-1} (1·81 cm^{-1}).

Molecular structure determinations

The prime interest in the determination of molecular moments of inertia is to be able to derive bond distances and interbond angles. The problem is a general one and is not confined to moments derived from measurements of pure rotational spectra; it is best considered in this context because of the wealth of accurate data that has been obtained from microwave measurements.

The simplest case is that of a diatomic molecule. For this the moment of inertia is given by

$$I = m_1 m_2 (m_1 + m_2)^{-1} r^2,$$

where m_1 and m_2 are the isotopic masses and r the internuclear distance. This is precisely the classical formula for a pair of mass points held rigidly apart at a distance r. No special quantum mechanical formulae apply to microscopic moments of inertia and the formulae of macroscopic dynamics apply.

The moment of inertia regularly obtained from the rotational constant is that of the ground vibrational state. This is usually implied by a zero subscript as in B_0, I_0 and r_0. For a diatomic molecule r_0 is strictly $[\langle r^{-2}\rangle_{\text{average}}]^{-1/2}$; the average is taken over the zero point vibrational motion. If the rotation–vibration interaction constant α is known then the equilibrium quantities B_e, I_e and r_e can be determined. r_e is the distance separating the nuclei when these are at the equilibrium distance and is more fundamental than r_0. The differences may reach 1 per cent. Using microwave moments of inertia, reliable α and isotopic masses, r_e may be obtained to 0·0002 Å, one of the limitations on accuracy being the precision with which Planck's constant, h, is known. The r_e are independent of the isotopic composition of the molecule to high accuracy, the discrepancies which might arise from the finite sizes of the nuclei being less than present experimental errors.

For some isotopes it is convenient to determine the isotopic mass from microwave rotational constant measurements assuming equilibrium distances to be unchanged. This determination is not affected by errors in h. In tables of isotopic masses the weight of the electrons required to form neutral atoms is included. If the electrons are not centred on the nuclei in question a correction must be applied. For instance with ^6LiBr and ^7LiBr the lithium isotope mass ratio from moments of inertia only agrees with the mass spectroscopic value if one electron is assumed to be transferred from the lithium to the bromine, as in the ion pair description Li^+Br^-.

Related comments apply to simple polyatomic molecules, but it is usually impossible to measure all the α so that B_e remains unknown. If there are as many isotopic species as unknown structural parameters, the moments of inertia may be used to determine average distances. Although values correct to 0·001 Å can be obtained in favourable cases, 0·005 Å is probably a more realistic value of the precision obtainable. Sometimes the microwave data prove insufficient for complete structural determination and assumption about some bond lengths or angles are incorporated so that the remainder may be evaluated. Sometimes the moment of inertia about the axis of a symmetric top may be obtained from infra-red or Raman spectroscopy. The errors which arise from the use of B_0 values instead of B_e are magnified when isotopic substitution is also used. The procedure outlined below for OCS can reduce these errors, but there are not always sufficient isotopic species known for its application. In some instances C–H and C–D distances determined from B_0 values appear to be appreciably different. This is essentially a feature introduced by the assumptions in the computation and does not imply that the equilibrium distances differ.

The problem of structural determination is of such importance that a simple case, that of carbon oxysulphide, is worth a more detailed discussion. Table 7.2 shows, amongst other quantities, the values of B_0

Table 7.2 *Microwave data for OCS*

	ISOTOPIC SPECIES	(B_0/MHz)	$(I_0/u\ \text{Å}^2)$	I (calc.)
I	$^{16}O^{12}C^{32}S$	6081·490	83·0820	82·9581
II	$^{18}O^{12}C^{32}S$	5704·83	88·5675	88·4403
III	$^{16}O^{14}C^{32}S$	6043·25	83·6077	83·4853
IV	$^{16}O^{12}C^{34}S$	5932·816	85·1640	85·0374

From I and II $z_O = -1{\cdot}6818$ Å $C{=}O = 1{\cdot}1612$
I and III $z_C = -0{\cdot}5206$ $C{=}S = 1{\cdot}5588$
I and IV $z_S = +1{\cdot}0382$
$m_O z_O + m_C z_C + m_S z_S = 0{\cdot}046$ u Å

I (calc.) values are obtained from the final C=O and C=S distances.

for four isotopic species. The molecule is linear and if z_C, z_O and z_S are the distances of the C, O and S atoms from the centre of mass of the first isotopic molecule, then its moment of inertia is given by

$$I = m_O z_O^2 + m_C z_O^2 + m_S z_S^2,$$

where the m are the atomic masses. The change of moment of inertia for a substitution of carbon of mass $m_C + \Delta m_C$ is given by

$$\Delta I = z_C^2[\Delta m_C - (\Delta m_C^2)(m_C + m_O + m_S + \Delta m_C)^{-1}]$$

and analogous expressions hold for other isotopic substitutions. The first component in the square brackets represents the main change, while the second corrects for the change of centre of mass on substitution. It is apparent that the use of a third isotope of carbon would not assist by providing a new equation containing z_O or z_S, but would merely provide confirmatory evidence for z_C.

The table indicates the results. The equation at its foot should read zero on its right-hand side since the origin of z is the centre of gravity and so it provides a check on the internal consistency of the method. So also do the moments of inertia of the last column recalculated from the values of z. If the moments of inertia of III and IV had not been available there would have been sufficient data to solve the problem. However, there is a favourable cancelling of the errors arising from the use of B_0 instead of B_e in the method described here.

Rotation in liquids and solids

In liquids the effective molecular collision frequency is very high and the rotational expressions for gases are no longer valid. The molecules obey essentially classical mechanics and no resonant rotational transitions are observed. However, this does not mean that the rotational absorption disappears. This can be seen by considering the Lorentz form of line shape for gases which gives absorption proportional to

$$\nu \Delta\nu / [(\nu_0 - \nu)^2 + (\Delta\nu)^2]$$

where ν_0 is the resonant frequency and $2\pi\Delta\nu$ the collision frequency. As $\nu_0 \to 0$ this expression becomes

$$\nu \Delta\nu / [\nu^2 + (\Delta\nu)^2] \equiv 2\pi\nu\tau / (1 + 4\pi^2\nu^2\tau^2).$$

$\tau \equiv (2\pi\Delta\nu)^{-1}$ would be the time between collisions in the gas phase, but must be taken as a characteristic time, the relaxation time, for condensed phases where collisions are not clearly defined.

The same form of frequency dependence for the loss can be obtained from the theory of dielectrics. This shows the real, ε', and the imaginary, ε'', part of the complex dielectric permittivity $\varepsilon = \varepsilon' - \mathrm{i}\varepsilon''$ as

$$\varepsilon' = \varepsilon_\infty + (\varepsilon_S - \varepsilon_\infty)/(1 + 4\pi^2\nu^2\tau^2)$$
$$\varepsilon'' = (\varepsilon_S - \varepsilon_\infty)2\pi\nu\tau/(1 + 4\pi^2\nu^2\tau^2).$$

Here ε_S is the static value of the dielectric permittivity (i.e. that at zero frequency) and ε_∞ is the high frequency value. ε_∞ strictly is the value at a frequency of about 10^{12} Hz (33 cm^{-1}) so that it is below absorption due to any vibrational transition. In the derivation from dielectric theory, τ is a time constant which is introduced to describe the slow rate of decay of the polarization P. If a condenser, filled with a dielectric and charged so that the dielectric polarization is P_0, is suddenly short circuited, then the polarization decays according to

$$P = P_0 \exp(-t/\tau).$$

The quantity $(\varepsilon_S - \varepsilon_\infty)$ is related to the molecular dipole moments present.

The exact relationship depends on the form of internal field assumed; if the Clausius–Mosotti relationship is valid then

$$(\varepsilon_S - \varepsilon_\infty) = (\varepsilon_S + 2\varepsilon_0)(\varepsilon_\infty + 2\varepsilon_0)Np^2/27kT\varepsilon_0^2$$

where N is the number of molecules per unit volume. Although not wholly appropriate for strongly polar liquids, this form is reasonable for dilute solutions in non-polar solvents and for weakly polar materials.

Apart from a relatively unimportant numerical factor which is usually close to unity and depends on the internal field, τ for dielectric loss, which is essentially the decay constant of the macroscopic polarization, is equal to the relaxation time, τ, of the rotational motion of individual molecules. τ may be thought of as the average time required by a molecule to rotate one radian, but it must be understood that this has a statistical interpretation. One individual molecule may rotate one radian in a longer or a shorter time. The rate of rotation $1/\tau$ rad s^{-1} is much slower than the instantaneous angular velocity since most molecules are frequently reversing their direction of rotation. An analogy is to diffusion velocity which is always less than the average of the instantaneous linear velocity.

Measurements of ε' and ε'' can be made by the standard techniques of electronic engineering. A.C. bridge circuits are used at low frequencies, transmission lines in the 100 MHz range and microwave cavities near 30 GHz.

Dilute solutions and weakly polar liquids obey the simple theory with

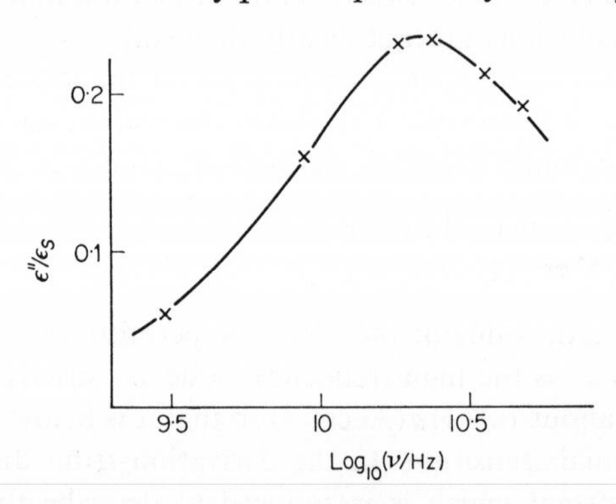

7·6 Dielectric loss of a molar solution of camphor in cyclohexane as a function of frequency. Full curve calculated for $\tau = 7 \times 10^{-12}$ s, $p = 1{\cdot}0 \times 10^{-29}$ m s A.

one relaxation time quite closely as indicated in Figure 7.6. This shows the experimental values of $\varepsilon''/\varepsilon_S$ (sometimes called tan δ) for solutions of camphor ($C_9H_{14}O$) in cyclohexane. ε_S is the solvent dielectric constant. Camphor is an essentially spherical molecule which rotates rather freely.

The full curve of Figure 7.6 shows the calculated value for $\tau = 7 \times 10^{-12}$ s and $p = 1{\cdot}0 \times 10^{-29}$ m s A.

The simple equation, often called the Debye equation, is less well obeyed by hydrogen bonded liquids, polymer solutions, solid polymers, glasses, etc., where wider rotational curves are common and may be interpreted in terms of a range of relaxation times. Highly crystalline solids seldom show rotational relaxation in the solid, although there are exceptions where there is appreciable rotation in the crystal.

No intramolecular information, apart from the dipole moment, is readily obtainable from dielectric loss measurements and the chief interest lies in the value of τ which is essentially an intermolecular feature. A selection of values are given in Table 7.3. Some idea of the

Table 7.3 *Some typical rotational relaxation times*

SUBSTANCE	(J/s)	(T/°C)
Chloroform ($CHCl_3$) in n-heptane	3×10^{-12}	25
Chloroform pure liquid	7×10^{-12}	25
Water (H_2O)	$9{\cdot}6 \times 10^{-12}$	20
Water	$3{\cdot}2 \times 10^{-12}$	75
Nitrobenzene ($C_6H_5NO_2$)	5×10^{-11}	20
Ethyl alcohol (C_2H_5OH)	$1{\cdot}4 \times 10^{-10}$	20
Glycerol (CH_2OH–$CHOH$–CH_2OH)	8×10^{-9}	0
Glycerol under 10 000 atmospheres	6×10^{-7}	0
Ice	4×10^{-6}	−5
Ice	2×10^{-3}	−45
Polymethyl acrylate (solid)	3×10^{-3}	25

forces opposing rotation can be obtained by treating $1/\tau$ as a rate constant. If the slope of a graph of ln τ against reciprocal temperature can be obtained this may be equated to E/R. Here E is the activation energy. The values of E vary from 5 kJ mol^{-1} for mobile systems up to 200 kJ mol^{-1} for some rigid polymers.

It is to be expected that similar forces will oppose rotation to those which control linear motion. Thus τ should vary with temperature, solute, etc., in a manner similar to diffusion and viscosity. This general statement agrees with experiment, but detailed relationships which have been suggested have proved inexact.

8 Infra-red spectroscopy

Introduction

There are two important distinct techniques for investigating the vibrational energy levels of a molecule, namely by measurement of its infra-red absorption spectrum and of its Raman scattering spectrum. The former is discussed in this chapter and Raman spectra in chapter 9; complete interpretation of either spectrum may require a knowledge of the other and a more comprehensive discussion of molecular vibrations is given in chapter 10.

For the present it is sufficient to appreciate that a molecule will have a number of fundamental vibrational frequencies and that each fundamental may be associated with absorption of radiation of its own frequency. There are $3N-6$ fundamental frequencies for a general molecule containing N atoms and for special molecules with degenerate frequencies, free internal rotations, etc., this number is only slightly reduced. The frequencies range from 100 to 4000 cm^{-1} and the absorption bands consequently lie in the infra-red region. Absorption may occur at non-fundamental frequencies if two or more fundamental modes of vibration are excited simultaneously, but the corresponding absorption is normally weak. The infra-red absorption spectrum of a complex molecule may thus show a large number of absorption bands of various intensities and with an irregular frequency pattern. Since the entire spectrum is characteristic of the molecule and no two species of molecule (except D and L enantiomorphs) have exactly the same absorption pattern, it is clear that the spectra can be used to identify compounds. Furthermore the identification can be made in a few minutes on a few milligrams of material and is still certain in the presence of small amounts of impurities, which may themselves be identified.

It is also possible to identify chemical groupings in a molecule from absorption at particular frequencies which is unlikely to arise from alternative sources. For example strong absorption at 2250 cm^{-1} indicates the presence of a cyanide group or a silicon-hydrogen bond or an unsaturated carbon-deuterium bond. The origin of the sample is likely to distinguish these possibilities. While such arguments from infra-red spectra do not provide complete certainty concerning the presence or absence of specific groupings, the indications obtained may greatly hasten structure determinations by other techniques.

Such applications of infra-red spectroscopy are so important that considerable effort has been directed to making spectrometers which are convenient, rapid, easy to use and relatively inexpensive. Such instruments are to be found in most chemistry research laboratories.

Experimental details

Basic spectrometer Many details influence the design of individual instruments, but the essential components are indicated in Figure 8.1.

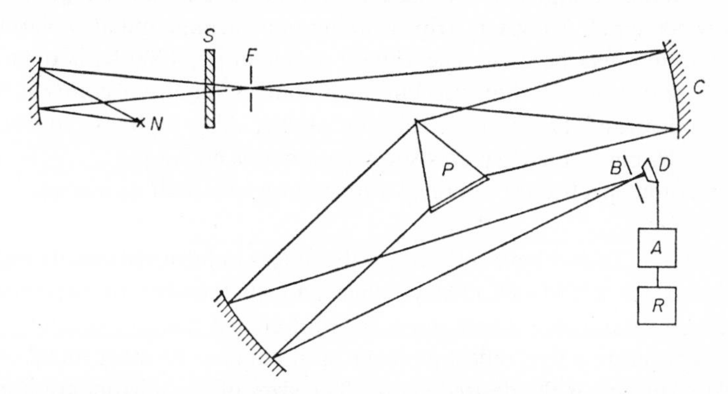

8.1 Arrangement of a simple infra-red spectrometer. N, Nernst source; S, sample; F, front slit; C, collimating mirror; P, prism; B, back slit; D, detector; A, amplifying circuits; R, chart recorder.

Radiation from a thermal source, N, is focused through the sample, S, and the front slit of the spectrometer, F. Hence the radiation is formed into a parallel beam by the collimating mirror, C, traverses the prism, P, and is refocused on to the back slit, B, and the detector, D. This is followed by a signal amplifier, A, and a recorder, R. In such an instrument the absorption is the ratio of the signal with the sample present to that when it is absent. The infra-red frequency is determined by the angular setting of the prism and may be varied by rotating the prism about a vertical axis.

Before indicating the forms of more sophisticated spectrometers it is appropriate to discuss individual component items in greater detail.

Radiation sources An infra-red source consists of a hot strip of material emitting all frequencies in accordance with the black body radiation law. The amount of energy at each frequency increases with the source temperature; the area of the source is unimportant provided the image of the source on the front slit is larger than this slit under all operating conditions. For most of the infra-red range the source energy per unit frequency interval is proportional to the fourth power of the frequency.

Electrical heating is most convenient and the commonest source material is a conducting mixture of rare earth oxides formed into a narrow tube and called a Nernst glower. This source runs at 2000 K but may require only 100 W and can run in air without special cooling. It has a negative temperature coefficient of resistance and requires external heating before it conducts sufficient current to remain alight. Thermal heating strips of silicon carbide run at lower temperatures, but are of larger diameter and may be used with wider slits; they are preferable below 500 cm^{-1}. Tungsten strips running in a vacuum and also platinum or rhodium may be used. The highest temperature, 4000 K, is obtained in the crater of a carbon arc but such sources require a constant feed mechanism and are difficult to keep stable. Also at low frequencies, below 100 cm^{-1}, mercury arcs show continuous emission.

Emission spectra are obtained with the sample itself as source.

Mirrors Lenses have serious disadvantages in infra-red spectrometers designed for a wide frequency range and all focusing of radiation is achieved by mirrors. These must be front coated and are usually made by evaporating a thin reflecting layer of aluminium or other metal on to a glass surface of the desired shape. The sizes of the mirrors are important if as much of the radiation as possible is to reach the detector. The collimating mirror must be as large as the face of the prism and of as short a focal length as is consistent with optical quality and design. This ensures that the radiation gathering power is a maximum since this is controlled by the aperture. All other mirrors should have at least as wide an aperture so that no radiation is lost at their edges. As camera owners will know, the aperture is written $f5$, for example, which implies that the focal length of the lens or mirror divided by its diameter is 5. The lower the f value the greater the radiation gathering power.

Slits The slits are normally two vertical straight edges, although some spectrometer designs call for slight curvature to reduce image aberrations. The distance between the slit jaws is variable and often the separate slits of a spectrometer are ganged together so that they can be operated from one control. If the slits are too wide spectral resolution is diminished, whereas if the slits are too narrow there is insufficient radiation reaching the detector to provide a satisfactory signal-to-noise ratio. A compromise must be made and, since the optimum is a function of frequency, the slits are varied during the recording of a spectrum if a large range of frequency is to be covered.

The slit width may be quoted as the actual separation of the jaws in mm, but a more useful quantity is the frequency spread at the given prism setting resulting from the finite size of the slits; this is the effective slit width and is given in cm^{-1}. Under common operating conditions a spectrometer will just resolve into two lines a pair of narrow absorption bands separated in frequency by the effective slit width.

Prisms The purpose of the prism is to select the required radiation frequency by virtue of the different deviation of the radiation of each frequency; radiation of neighbouring, unwanted frequencies are focused on the opaque jaws and surround of the back slit. The choice of prism material is influenced by the frequency at which measurements are required. It is essential that the prism be transparent to the required radiation. It is also desirable that the system should have resolution which is adequate for resolving neighbouring bands. According to the exact conditions, the resolution may be limited by the inherent resolution of the prism, by lack of energy reaching the detector at narrow effective slit widths or by a combination of these causes. Under all conditions the following factors are favourable for high resolution. (i) A high rate of change of refractive index with frequency, $dn/d\nu$; (ii) an increase of the size of the prism, i.e. of refracting angle, base length and height; (iii) optical layouts which permit multiple passage of the radiation through the prism. The commonest of these is the Littrow arrangement with a plane mirror at the back of the prism, but other arrangements with four or more traversals have been devised; (iv) a narrow slit in relation to the focal length of the collimating mirror. Any feature, such as improved source, improved detector, high mirror reflectivity, etc., is valuable, if it leads to more effective energy utilization.

Table 8.1 shows a list of useful prism materials. Below the frequency quoted the materials are useless because of very strong absorption. At higher frequencies $dn/d\nu$ falls gradually and is too low to be very useful at frequencies about five times the value quoted. Each material is therefore most effective near its lower frequency limit. But the prisms may be expensive and cannot always be exchanged rapidly and so they are often used above their most favourable range if the highest possible resolution is not required. General purpose spectrometers are commonly

Table 8.1 *Useful prism materials for the infra-red*

Above 200 cm^{-1}	Caesium iodide
,, 250 ,,	Caesium bromide
,, 380 ,,	Potassium bromide
,, 670 ,,	Sodium chloride
,, 1100 ,,	Calcium fluoride
,, 1600 ,,	Lithium fluoride
,, 2800 ,,	Fused silica
,, 5000 ,,	Flint glass

fitted with a sodium chloride prism and cover the range 700–4000 cm^{-1}. The prisms may be 10–12 cm in height with other dimensions in proportion. They are mostly grown from the melt as single crystals specially for use in infra-red spectrometers.

Windows Transparent windows are frequently required for sample cells, detectors, etc., and these too must be made of transparent materials. Even though $dn/d\nu$ is unimportant the list of prism materials includes most important window materials as well. For cell windows chemical inertness may be important and this raises special difficulties with aqueous solutions. The first four salts in Table 8.1 are soluble in water and at low frequencies the best material for containing water is a mixture of thallium bromide and iodide known as 'KRS 5'. More easily obtainable are silver chloride sheets which are transparent above 400 cm^{-1}. Unfortunately they are rather soft and sensitive to visible light. Polyethylene may also be used below 650 cm^{-1}.

Gratings The prism may be replaced by a reflecting grating. This form of dispersive component has the advantage of improved resolution, wider range and, at least at low frequencies, reduced cost. Indeed grating spectrometers are very rapidly becoming more common. Gratings for use in the infra-red may be up to 20 cm square and have from 300 to 3000 lines per cm. If the incident light is at normal incidence, radiation of frequency ν cm^{-1} is reflected at an angle θ to the normal given by

$$\sin\theta = n/\nu d,$$

where n is the order and d the grating spacing in cm between adjacent rulings.

The lines may be shaped so that the reflected radiation is concentrated in a given direction and the grating is then said to be blazed for this direction or the corresponding frequency. Blazing reduces the wastage of energy in unwanted orders with corresponding increase of signal.

Many modern gratings are plastic replicas of an original ruling. The thin replica is mounted on a glass optical flat and coated with evaporated metal to obtain the reflecting front surface.

The chief disadvantage of a grating is that while it efficiently separates radiation of the desired frequency from radiation of adjacent frequency, it does not separate it from radiation of double or higher multiple frequency. Such radiations satisfy the basic grating equation with $n = 2, 3, 4$, etc. This radiation is referred to as radiation of the second, third, fourth, etc., order. These higher orders may be used intentionally as the resolution, for a given grating and frequency, is proportional to the order provided blaze efficiencies are not unfavourable. Radiation of unwanted orders may be efficiently removed by adding a small prism spectrometer, which is often placed before the entrance slit. Accurate correlation of the rotation of the prism and of the grating is required if both are to pass the same frequency simultaneously. Alternatively radiation filters may be used, but suitable materials are not available for all regions.

Michelson interferometers Mention should be made of the use of interferometric methods which are especially valuable when energy limitations are severe, as they are in the extreme infra-red (10 to 200 cm^{-1}). The radiation to be analysed is split into two beams which are finally recombined. A moving mirror continuously varies the path difference between the beams. At any instant the detector will find itself at an interference maximum for some frequencies and at a null for others. Its total output is thus a complicated, but known, function of the energy of the source as modified by the sample. The resulting interferogram, that is the signal as a function of the position of the moving mirror, contains the spectrographic information which may be extracted therefrom via a computer or suitable wave analyser.

The energy advantage comes from a larger aperture, the use of a circular rather than slit-shaped entrance and detector, and particularly because energy of each frequency may reach the detector throughout the operating time. This is in contrast to a spectrometer where the energy is all stopped by the back slit, except for the narrow range being currently observed. The optimum resolution of such an interferometer is the reciprocal of the distance of travel of the mirror which may be several centimetres. If the sample is placed in only one of the beams, its refractive index, as well as its absorption, may be obtained as a function of frequency.

Thermal detectors The common infra-red detectors are essentially rapid, low heat capacity thermometers in which the temperature rise due to the radiation falling on them is registered in convenient electrical form. They must be sensitive to very small heat inputs and return to their natural ambient condition when the radiation is removed. The surface receiving the radiation must be blackened to be an efficient absorber.

One form is a small, 2 mm × 0·2 mm, radiation receiver mounted in a vacuum and associated with a thermocouple.

Resistance thermometers may also be used in which case the active material, chosen for its large temperature coefficient of resistance, is itself the receiving element. A steady electrical current is supplied and the voltage changes amplified. The name bolometer detector is applied to such devices.

A third style of thermal detector is the pneumatic detector or Golay cell. In this the radiation heats a gas sample whose consequent expansion distorts a thin containing wall. This distortion is measured by using the distorting membrane as a reflecting mirror with an auxiliary light source and photocell.

All three devices are extremely delicate and require careful handling. They are of comparable sensitivity and work at room temperature. The minimum detectable energy is about 10^{-10} W for 1 Hz bandwidth at the recorder. Superconducting bolometers and carbon resistances at liquid helium temperatures may improve on this figure, which is essentially independent of radiation wavelength.

Quantum detectors The minimum detectable energy may be reduced below that for thermal detectors by the use of photoelectric and especially photoconductive devices. These are essentially semiconductors, whose electrons may be excited to the low-lying conduction levels by infra-red quanta. The consequent change of conduction is observed. Individual detectors have a narrower frequency range for efficient operation, lead sulphide being useful near 3000 cm^{-1}, lead selenide and indium antimonide near 1800 cm^{-1}, and doped germanium crystals show promise at lower frequencies down to 200 cm^{-1}.

Radiation chopping Since it is the signal due only to the source radiation which is to be measured, provision must be made for discounting the signal due to the temperature of the room and similar irrelevant sources. In the simplest hand-operated systems the radiation is cut off by an opaque mechanical shutter and the difference in output meter reading is the desired signal. For recording systems a rotating sector is used and the amplitude of the alternating component at the sector speed is proportional to the required signal. The chopper speed must be slower than the inherent time of response of the detector so that frequencies from 3 to 30 Hz are used with thermal detectors, whereas

800 Hz is more convenient for quantum detectors. Phase sensitive detection is regularly employed and the overall bandwidth of the system is that of the final electrical smoothing.

Recording spectrometers In view of the widespread application of infra-red absorption spectra, much effort has gone into improving the overall speed and convenience of spectrometers. Improvement in detector design by about 1940 made it possible to couple the output to a pen recorder and to rotate the prism turntable by an electric motor so that a continuous record of radiation energy against frequency was obtained. Diminution of energy indicated absorption bands of the sample, but in view of the variation of source energy with frequency a separate blank record had to be run in the absence of the sample before it was possible to obtain quantitative results. Absorption by atmospheric water vapour and carbon dioxide also interferes with the record and is best removed by flushing with nitrogen gas or evacuating the whole spectrometer.

Various designs of double beam spectrometer have been built to overcome these difficulties. These spectrometers incorporate two radiation paths for part of the optical system. One traverses the sample and the other a matched path with an empty cell or other comparison unit. The resulting signals from each path are appropriately sorted and the ratio is recorded by the pen recorder. The ratio, unlike the signals themselves, is essentially independent of the slit widths and these may be gradually varied according to a preset programme so that a compromise between high resolution and high signal-to-noise ratio is maintained throughout the extended frequency interval of a single record. If carefully shaped cams are used to control the prism or grating rotation, preprinted recorder paper may be used, so that an acceptable permanent record is obtained directly from the instrument.

Various methods of extracting the ratio of the signals in the two paths have been used, but the most effective is to equalize them by inserting a wedge or comb into the beam which does not contain the sample. The exact position of the comb, when the beams are equalized, indicates the absorption by the sample and a rapid adjustment to equality can be arranged by a servo system working from the detector output. The chopping system incorporates a moving mirror so that the radiation from the two paths alternately reaches the detector. If the system is balanced with the comb in its correct position there is no alternating signal at the detector; if the comb is inexactly inserted, the phase of the alternating output can be used to control the direction of movement of the comb.

Some spectrometers incorporate features which allow scale expansion to give full scale deflection for 10 per cent or 1 per cent absorption, variable scanning rate so that time is not wasted on transparent regions, automatic stop, start and repeat buttons, distant slave recorders, automatic prism or detector changes, etc. Oscilloscope presentation, albeit with necessarily reduced resolution, is sometimes used particularly in the study of transient species.

Samples A very wide range of specimen conditions can be used for infra-red absorption. Gases and volatile liquids are normally measured in gas cells about 10 cm long at a pressure of a few k Pa. Special cells with multiple internal reflection have effective path lengths up to 40 m and are useful for materials with low vapour pressures. Sometimes total gas pressures up to several atmospheres are used to study weak transitions.

Liquids can be measured in thin cells between windows of sodium chloride about 0·1 mm apart. Thinner or thicker cells are usually available, while non-volatile, viscous liquids can be smeared over one plate. Special cells are made for small quantities down to 10^{-9} m^3 and also for temperatures above and below room temperature.

Solid powders present more difficulty because of the loss of radiation by scattering. This loss is approximately proportional to the fourth power of the frequency and may not be serious below 1000 cm^{-1}. The loss is also proportional to the fourth power of the refractive index difference between the solid particles and the surrounding medium. Therefore scattering is reduced if the surrounding air is replaced by any transparent liquid. Below 1300 cm^{-1} a liquid paraffin—often given its American trade name Nujol—is efficient and above this a fully halogenated liquid such as hexachlorbutadiene, $CCl_2=CCl-CCl=CCl_2$, is best. These absorb in other regions so that only a partial spectrum is obtained, although the regions obscured by paraffin may not be too important. An alternative technique has been developed whereby the air is exchanged for potassium bromide. A 1 per cent mixture of sample in the salt is finely ground and compressed in a vacuum to form a transparent solid disc with the sample embedded in a potassium bromide matrix. Some solids can be compressed or melted to give transparent films and polymers may be cast from solvents in the form of thin films. Single crystals may also be used; these are often studied in conjunction with a small reflecting microscope, which is valuable for other small specimens.

Solids and liquids may also be examined in solution. Since no solvent is transparent for the entire region a series of solvents are required. Carbon disulphide, CS_2, is very useful outside the region 1400–1700 cm^{-1} and tetrachloroethylene, $CCl_2=CCl_2$, is a useful complementary solvent which is relatively transparent here. More polar materials have more strong absorption bands and are less valuable, although this disadvantage is partially offset by the greater solubility of most materials. If the solvent bands are not too strong it is possible to introduce pure solvent in the comparison beam of a double beam spectrometer so that the solvent absorption is eliminated from the recorded spectrum. Solution spectra are convenient to measure and easy to reproduce and are therefore favoured for quantitative analysis, when a wide frequency range is less important.

Material identification

Identification of materials by means of their infra-red spectra is so certain and convenient that it is replacing older methods based on melting-points, refractive indices, etc. Its chief advantages lie in the certainty of the conclusion even with slightly impure materials. The major disadvantage is the amount of space required for displaying the full details of a spectrum. As a consequence scientific journals will not normally print the spectrum if the sole reason is to enable workers in other laboratories to identify the compound when they make it for themselves. This difficulty is being overcome by special publications, usually in the form of loose sheets or loose charts, devoted specifically to such spectroscopic data. Also most research laboratories have extensive files on limited classes of compounds of special interest to that laboratory and these are available to all workers in the same institution. Copies of spectra are exchanged by post more easily than samples for mixed melting-points and questions of deterioration of specimens are irrelevant. There will always be small differences arising from the use of different spectrometers and different conditions of sample preparation but these are seldom troublesome. This use of infra-red spectra has led to their being called, with considerable justification, the fingerprints of chemical compounds.

Figure 8.2(c) shows part of the spectrum of a product believed to consist of equal parts of 2-methyl dioxalane and 4-methyl dioxalane, whose spectra are given in Figures 8.2(a) and (b) respectively. Immediate inspection shows that (c) resembles (b) much more closely than (a) and the product must contain approximately 70 per cent of 4-methyl dioxalane. The presence of about 20 per cent of 2-methyl dioxalane is indicated and the peaks marked by arrows near 1118, 1024, 863 and 692 cm^{-1} are certainly due to this component. However, two new peaks, marked by crosses, near 1060 and 780 cm^{-1} show a third component to be present, as does the increased absorption at 835 cm^{-1} relative to that at 825 cm^{-1}. 2,2-dimethyl dioxalane has strong peaks at 1061, 837 and 782 cm^{-1} and its other strong bands in this region would be masked by bands of

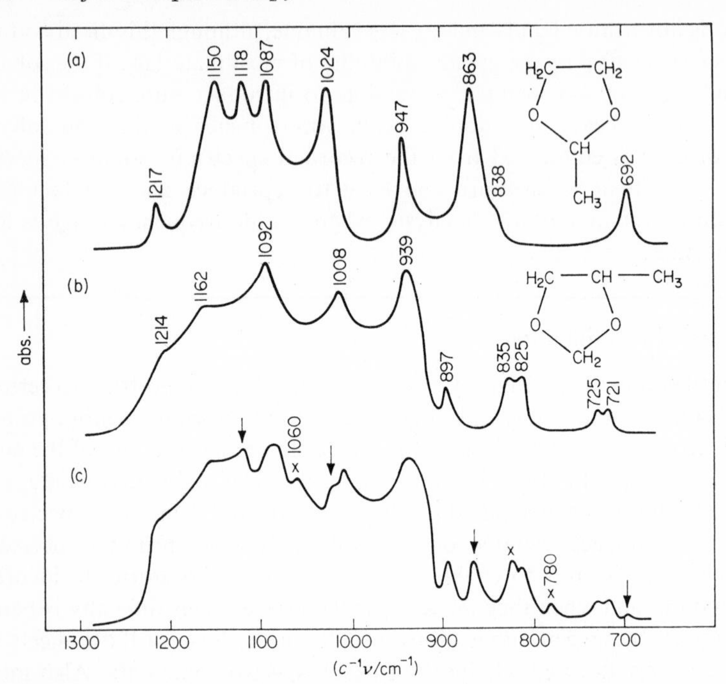

8.2 Infra-red absorption spectrum of (a) 2-methyl dioxalane, (b) 4-methyl dioxalane, (c) mixture with trace of impurity.

4-methyl dioxalane. 2,2-dimethyl dioxalane is consequently strongly suspected of being the impurity.

Quantitative analysis

Since absorption is a function of the amount of material in the radiation path, it is possible to use infra-red spectra for quantitative analysis.

Consider a concentration, c, of a material with an absorption band at frequency, ν, in a transparent solvent contained in a cell of path length l. Let the radiation of frequency ν travelling parallel to the x direction have intensity $I_\nu(x)$ at any point and in particular have the value $I_\nu(0)$ at the front surface of the solution. See Figure 8.3. In traversing an increment of cell, dx, there is a reduction of $I_\nu(x)$ which is proportional to $I_\nu(x)$ and the distance travelled dx. For solutions without specific concentration dependent effects such as dimerization, the loss of radiation intensity is proportional to the concentration of absorbing material present. If K_ν is the proportionality constant then these features are expressed by

$$-\mathrm{d}I_\nu(x) = K_\nu I_\nu(x) c \,\mathrm{d}x.$$

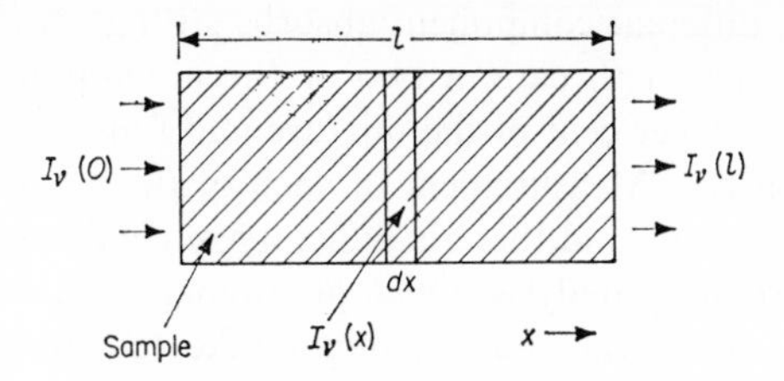

8.3 Arrangement considered in text for deriving quantitative absorption law.

This integrates to give

$$\left| -\ln I_\nu(x) \right|_{I_\nu(0)}^{I_\nu(l)} = \left| K_\nu c x \right|_0^l$$

or $\ln [I_\nu(0)/I_\nu(l)] = K_\nu cl$

or $I_\nu(l)/I_\nu(0) = \exp(-K_\nu cl)$.

The left-hand side of this last expression is the fraction of the radiation transmitted by the solution. Lambert first suggested this form of dependence on cell length and Beer the dependence on concentration and the expression is sometimes known as the Beer–Lambert law.

When c is in mol/litre, l in cm and logarithms to the base 10 are used, the proportionality constant is often called the molecular extinction coefficient and denoted by ε. That is $\varepsilon = K/2{\cdot}303$. The quantity $\log_{10} I(0)/I(l)$, is called the optical density and is provided as a direct scale reading on some spectrometers. The strongest infra-red bands in liquids or solids have $\varepsilon \sim 1000$ at their maximum and anything with ε above 50 would be strong, while weak bands have $\varepsilon_{max} \sim 1$.

For mixtures of absorbing materials the incremental absorption is additive in the absence of chemical interactions and so

$$-\mathrm{d}I = (\Sigma_i K_i c_i) I(x)\, \mathrm{d}x$$

and $\ln [I(0)/I(l)] = (\Sigma_i K_i c_i) l$

at each wavelength. l may be presumed known and determination of the transmission is the main measurement. $\Sigma_i K_i c_i$ is thus obtained and provided the K_i are known from measurements on pure materials, an equation in the unknown c_i is obtained. Measurements at a new frequency give an independent equation in the c_i. If as many frequencies are studied as there are unknown concentrations, the resulting set of simultaneous equations may be solved to obtain these concentrations.

The accuracy of this analytical method depends considerably on the choice of frequencies for observation. Ideally each frequency should be one at which a different component absorbs strongly and the others are nearly transparent. With care, a two or three component mixture can be analysed to ± 1 per cent of the mixture and four or five components with lesser accuracy. Measurement of each of the K_i at each frequency may be rather tedious and there is difficulty in choosing the best conditions for a particular analysis. Once preliminary calibration data have been obtained, a typical analysis might take ten minutes for sample preparation, ten minutes for obtaining the spectra and ten minutes for computing the results. Advantages in infra-red analysis include its use of small samples, recoverable if required, its ability to analyse closely related chemical materials, and its adaptability to non-volatile materials. The method is comparatively cheap and rapid. Disadvantages are indifferent accuracy, the paucity of suitable solvents and the difficulty of transferring exact K_i between different spectrometers.

Qualitative group analysis

Although the molecular vibrations associated with infra-red absorption are strictly vibrations of the whole molecule, it often happens that the distortion is largely confined to one section of the molecule. If this section recurs in different molecules, the associated infra-red absorption will reappear at essentially the same frequency and approximately the same intensity. This happens so reliably that the presence or absence of absorption at a particular frequency can be used to infer the presence or absence of the chemical grouping concerned. The absence of groups is more reliably determined, since absorption in a certain region may arise from some unusual combination of groupings which would separately be transparent.

Such frequencies associated with chemical groupings are called characteristic group frequencies. Collections of such have been made and these represent the accumulated experience of many investigators. Table 8.2 is an abbreviated form of such a table with some entries applicable to aliphatic compounds containing only C, H and O.

Although the exact directions of atomic motions during a vibration are only obtainable with difficulty, approximate vibrational forms are often helpful. Motion of two atoms forming a bond along the direction of that bond is described as a stretching motion, whereas perpendicular motions are deformations. These and other descriptions are indicated in Figure 8.4. In this, arrows represent the directions of atomic motions in the plane of the paper, while + and − describe motions out of this plane upwards and downwards respectively. In the opposite phase of the motion all the arrows and signs are reversed.

Table 8.2 *A selection of characteristic infra-red absorption frequencies*

		$(c^{-1}\nu/\mathrm{cm}^{-1})$	INTENSITY
Hydroxyl	O–H stretch	3590–3650*	s
Olefinic	=C–H stretch	3000–3100	w
Saturated	C–H stretch	2850–2970	w
Carbonyl	C=O stretch	1680–1780	vvs
Unsaturated	C=C stretch	1620–1680	w
Methylene	$>CH_2$ scissor deformation	1460	m
Methyl	$C{-}CH_3$ symmetrical C–H deformation	1375	w
Esters	C–O stretch	1150–1300	vs
Vinyl group	$-CH{=}CH_2$ hydrogen out-of-plane bends	985–995 & 905–915	m m
Vinylidene	$C{=}CH_2$,, ,,	885–895	m
Trans	–CH=CH– ,, ,,	960–970	m
Lone hydrogen	>C=CH– ,, ,,	790–840	m
$-(CH_2)_x-(x>4)$	hydrogen rock	720	m

w = weak, m = medium, s = strong, v = very
* Lower if hydrogen bond present.

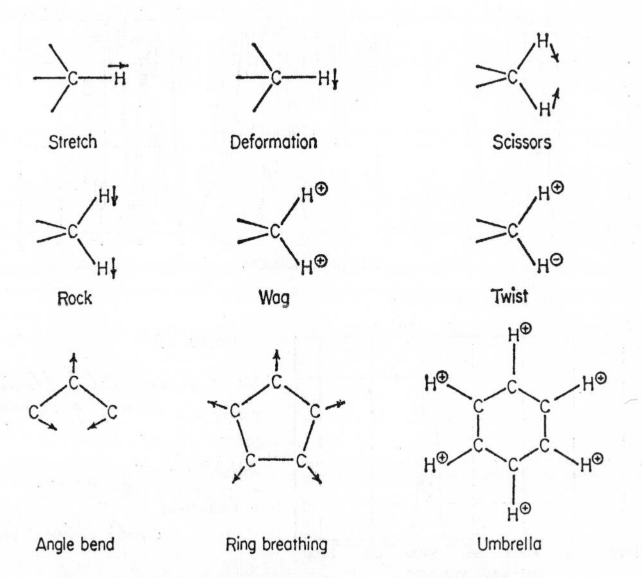

8.4 Approximate motions in some simple vibrational forms. + and − refer to motions perpendicular to the plane of the page.

Structure determination—an example

Since the assignment of chemical formulae to materials of previously uncertain structure forms a considerable proportion of the use of infra-red spectroscopy, an example given in some detail may be appropriate. Figure 8.5 shows the spectrum from 650–4000 cm^{-1} of a material,

$CF_3COCH_2CH_2COOH$ I

$(CF_3COCH_2CH_2CO)_2O$ II

$F_3C—C(OH)—CH_2$, O, CH_2, C=O (ring) III

$F_3C—C{=}CH$, O, CH_2, C=O (ring) IV

$F_3C(CH_2CH_2COOH)C{=}C(CH_2—CO—CF_3)(COOH)$ V (Cis and trans isomers)

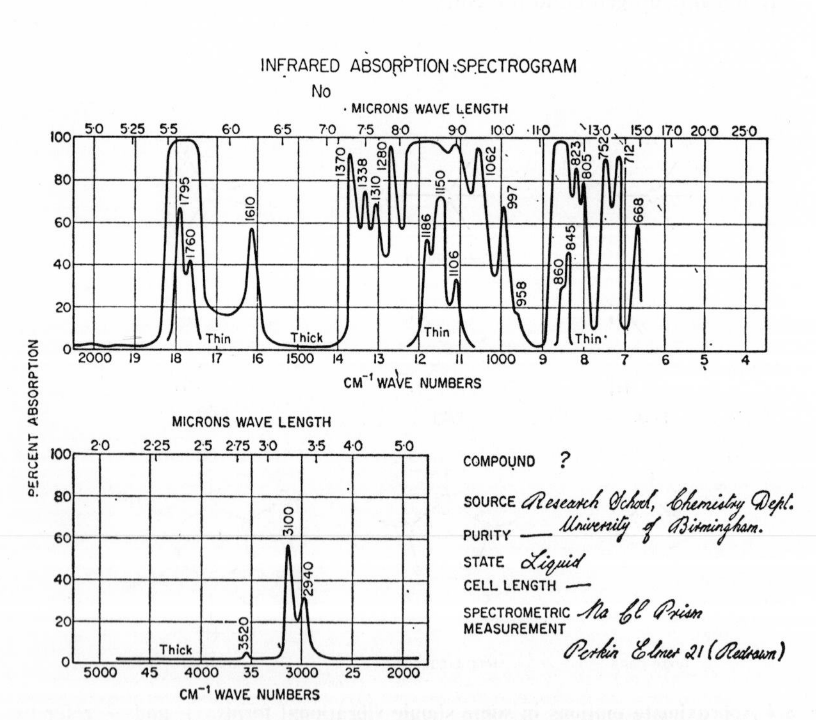

8.5 Spectrum of unknown product from 5,5,5-trifluor-laevulinic acid, I.

B.Pt. 179°C, obtained by the distillation of 5,5,5-trifluor-laevulinic acid, I, from phosphorus pentoxide. The spectrum is that of liquid layers between sodium chloride plates and two thicknesses are shown to bring out the details. From the method of preparation the most plausible structures are II–V of which the last would exist in cis and trans forms. The absorption peaks may be considered individually. The information quoted here is based on *The Infra-red Spectra of Complex Molecules* by L. J. Bellamy, which is the standard source for characteristic frequencies and the references to the original literature can be found therein.

Absorption at 3520 cm^{-1} This peak is possibly due to the stretching vibration of an O–H group. The intensity here is fairly weak and the –OH group is more likely to be due to an impurity than to the main species.

Absorption at 3100 cm^{-1} The band at 3100 cm^{-1} is one of the most significant in the whole spectrum. It must be due to the stretching of an olefinic or aromatic C–H group. IV is the only suggested structure which contains such a feature.

Absorption at 2940 cm^{-1} This must be due to stretching vibrations of an aliphatic C–H group. For a $>CH_2$ group there is normally a second C–H stretch absorption band near 2850 cm^{-1}; this is nearly as strong in hydrocarbons but is of reduced intensity if the $>CH_2$ group is adjacent to a $>C=O$ group. Only a broad indistinct absorption is shown in Figure 8.5 which is more consistent with IV than structures containing the $-CH_2-CH_2-$ grouping.

Absorption at 1795 cm^{-1} The very strong band at 1795 cm^{-1} is certainly due to a $>C=O$ stretching vibration, but it is higher in frequency than most carbonyl bands which are nearer 1730 cm^{-1}. Features which are known to raise the $C=O$ frequency include its being exocyclic to a small strained ring and the presence of a double bond adjacent to the ester oxygen as in vinyl acetate, $CH_3CO-O-CH=CH_2$. Both these features occur in $\beta\gamma$ unsaturated γ lactones, of which IV is an example, and about 1800 cm^{-1} is the regular value for such lactones.

Absorption at 1760 cm^{-1} The moderate band at 1760 cm^{-1} could be a second $>C=O$ frequency, which is not expected for IV. It may indicate a trace of impurity, but there is another possible explanation, namely that it is not a fundamental frequency but is a combination band which has acquired intensity at the expense of the very strong fundamental.

Absorption at 1610 cm^{-1} The value 1610 cm^{-1} is that to be expected for a $>C=C<$ valency stretching vibration.

Remaining bands Lower frequency bands are always more difficult to assign with certainty and are in general less characteristic. The ranges 1250–1400 and 1100–1200 cm^{-1} cover the C–F stretching frequencies of the $-CF_3$ group which are usually, as here, very strong. Also the ring frequencies due to C–C and especially C–O stretching will be strong and lie in the range 1000–1250 cm^{-1} for structure IV. The band at 845 cm^{-1} due to the C–H out-of-plane deformation is also fairly strong and, if correctly identified, of rather higher frequency than in hydrocarbons where 790–840 cm^{-1} is a more usual position.

Transmission regions Often the absence of bands may provide important evidence. In the present instance the absence of a strong –OH band near 3500 cm^{-1} argues against structure III. The absence of much absorption at 2850 cm^{-1} is against structures II, III and V with several $>CH_2$ groups each. Carboxylic acids, particularly in dimeric or other associated forms, absorb strongly in a wide region, 2500–2700 cm^{-1}, where this substance is transparent. If II were the correct structure a strong anhydride band near 1810 cm^{-1} would be expected, while V should show $>C=O$ absorption near 1720 cm^{-1} due to the –COOH groups.

Discussion The available evidence firmly favours structure IV. This means that this structure may reasonably be accepted as the result of about an hour's work. Certainly any question of proof by synthesis would start with this structure and chemical tests can be devised with special reference to IV.

9 Raman spectra

Principles

An important means of obtaining molecular vibration frequencies is from Raman spectra. The frequencies so obtained are for the greater part the same as those obtainable from the infra-red spectrum of the molecule, since they relate to the very same molecular vibrations. However, it may happen, especially with symmetrical molecules, that some prominent frequencies in the Raman spectrum are absent from the infra-red spectrum and *vice versa*. Indeed for molecules which possess a centre of symmetry there is a mutual exclusion rule which states that no frequency should appear in both spectra.

The basic process of the Raman effect is the removal of energy by the molecule from the quanta of a strong incident beam of light. The light, depleted of this energy and consequently of lower frequency, is scattered in all directions. The frequency spectrum of the scattered light is analysed by the spectrometer and, for monochromatic incident light, there is one new frequency observed for each active molecular vibrational frequency.

The energy balance for such a process requires that the energy retained by the molecule equals the energy of the incident quantum minus the energy of the scattered quantum. Thus the basic equation is

$$\nu_j = \nu_0 - \nu_R.$$

Here ν_j is the frequency of one of the active molecular vibrations, ν_0 is the frequency of the incident monochromatic light and ν_R the frequency of the scattered or Raman radiation. Such studies are named after C. V. Raman, the Indian scientist who first observed the effect in 1928, following its prediction by Smekal.

Light of unchanged frequency, ν_0, is also scattered by the sample and if the scattered light is analysed by a photographic spectrometer the frequency separations between this line, called the Rayleigh line, and the

Raman lines are equal to the molecular frequencies. Regularly the Raman lines are at lower frequencies, but lines may be observed on the high-frequency side of the exciting line. Such lines are called Stokes and anti-Stokes lines respectively. The lines of higher frequency result from the process whereby an incident quantum extracts energy from a molecule which is initially in a vibrationally excited state.

The Raman lines are always extremely weak with respect to the incident light intensity. The exact intensities depend on the molecular polarizability and more particularly the derivatives of the polarizability with respect to the molecular distortions. Just the same factors govern the intensity of the anti-Stokes Raman lines, but in conditions of thermal equilibrium these are weaker by a Boltzmann population factor, $\exp - (h\nu_j/kT)$, since only initially excited molecules contribute. Such lines are easiest to observe for low ν_j when this factor is not too unfavourable.

Experimental

Figure 9.1 shows schematically a typical photographic spectrometer. The light source is usually an intense mercury discharge tube whose blue line

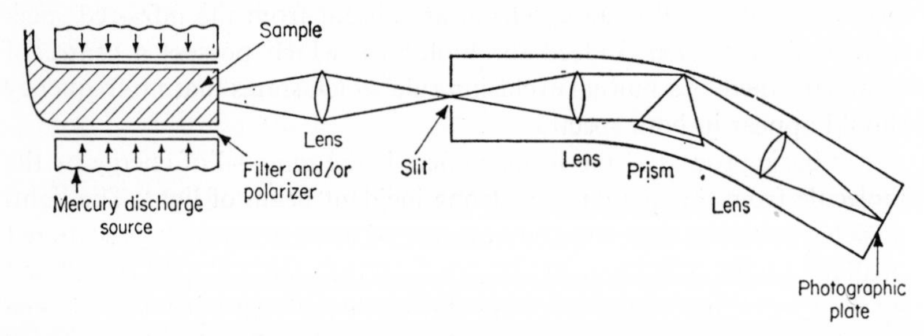

9.1 Simple photographic Raman spectrometer.

at 4358 Å is most useful. The source surrounds the sample tube and the geometry of the tubes and associated reflectors is such that as much radiation as possible enters the sample. Filter solutions, and cooling jackets may be placed between the source and sample. Liquids are placed in tubes with flat end windows through which emerges the scattered light, but not the direct incident light. This scattered radiation is focused on the front slit of the spectrometer by a lens. The spectrometer needs to be of high quality and normally has glass prisms and lenses with a photographic plate for detector. A wide aperture is desirable to reduce exposure times which may be several hours. Good dispersion is required to enable accurate frequency measurements to be made. In this arrangement polarization measurements are made by wrapping a piece of polaroid around the tube, firstly using a sheet transmitting light with its electric vector parallel to the cylinder axis and secondly with a perpendicular orientation. Other conditions being equal the ratio of the intensities of the scattered lines in these cases is the depolarization ratio, ρ. For fully polarized lines $\rho = 0$, for partially polarized lines $0 < \rho < 6/7$, and for depolarized lines $\rho = 6/7$. Different lines in a given spectrum will have different values of ρ. For some other geometrical arrangements of the source a Nicol prism may be inserted between the sample and the spectrometer for polarization measurements. It is very probable that laser sources will be increasingly used for Raman work and these involve consequential changes of geometry of the arrangements before the entrance slit.

Some modern spectrometers replace the photographic plate with a photocell or photomultiplier device so that after amplification the signal is displayed on a pen recorder. Successive frequencies are brought to the detector by rotating the prism or associated mirror as infra-red spectrometers. Recording spectrometers are more convenient and more accurate for intensity measurements, but are not as suitable for very weak lines.

Normal samples are 8–10 cm^3 of liquid, though the most modern spectrometers can obtain satisfactory spectra with only 0·3 cm^3 of a pure liquid. Since the amount of scattering is proportional to the amount of material, gases present difficulties. With special facilities, including multiple reflection cells filled to several atmospheres pressure, gases can be used successfully. Solids scatter badly from their surfaces and this light, though of unchanged frequency, may cause a fogging of the photographic plate at the sides of the Rayleigh line at ν_0. Consequently, if possible, solid materials are measured in solution. In contrast to the infra-red, water is a good solvent for Raman work. Single crystals can be used if large enough and with special source geometries even powders can be studied successfully. The Rayleigh line may be reduced in intensity by a filter between the specimen and the spectrometer.

Coloured samples are best measured with a lower exciting frequency, where their absorption is less. If light which is absorbed is used for exciting radiation some of the assumptions of the theory are not justified and its details must be modified especially as regards the intensities.

Figure 9.2 shows the Raman spectrum of 4-methyl pyridine observed with a recording spectrometer. The strong scattering at the origin on the right corresponds to the exciting line, Hg 4358 Å, which would obscure shifts below 100 cm^{-1} unless they were very intense. About twenty shifts to low frequency (Stokes lines) are observed of which the strongest at 994 cm^{-1} may be attributed to a ring breathing motion. For spectrum (a) the polaroid wrapped around the sample tube transmitted light with its

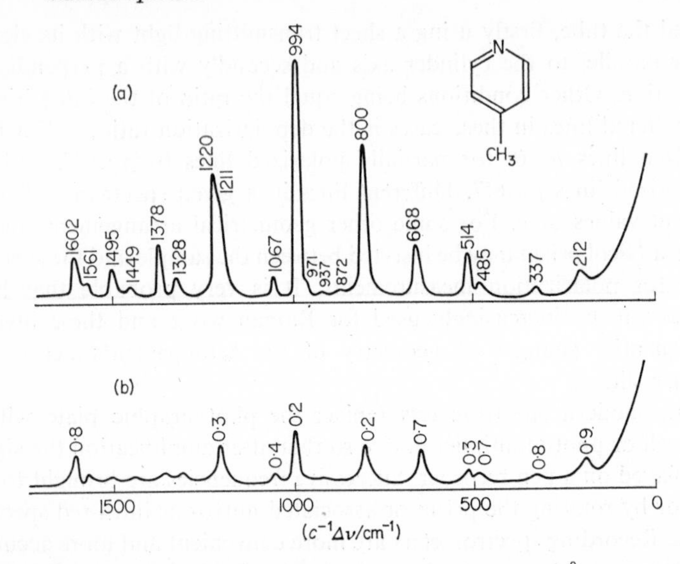

9.2 Raman spectrum of 4-methyl pyridine excited by Hg 4358 Å line. (a) Incident light with its electric vector perpendicular to scattered light, (b) parallel. Numbers in (a) cm^{-1} shift, (b) intensity ratio.

electric vector perpendicular to the tube axis, whereas for (b) the electric vector of the exciting light was parallel to the tube axis and hence also to the direction of travel of the scattered light entering the spectrometer. Since the electric vector of the emergent light from the Raman tube must be perpendicular to its direction of travel, spectrum (b) corresponds to depolarization on scattering and is weaker. For some lines, for example that shifted by 668 cm^{-1}, the relative strengths are 6 : 7 and the scattering is due to depolarizing vibrations, but for other lines the intensity is much reduced indicating that a proportion of the scattered light retains its polarization. As exemplified by the shifts of 800, 994 and 1220 cm^{-1} in this case, it is commonly the most intense lines which are partially polarized.

General applications

Most of the applications discussed for infra-red spectroscopy in chapter 7 are equally applicable for Raman spectroscopy although this is much less often used. Sample identification is just as easy and may be as rapid with photoelectric recording, though the trouble of photographic techniques is a deterrent. So also is the amount of sample required and the difficulty with insoluble solids. Quantitative analysis is relatively straightforward; the intensity of any Raman line is proportional to the concentration of the substance which gives rise to it. Small corrections need to be made for refractive index changes and the cell geometry must be held unchanged.

Group analysis is also possible and the infra-red frequency tables apply generally to the Raman spectrum although the intensities may be very different. There are a few groups which give rise to strong Raman lines and very weak infra-red absorption. Fully substituted ethylenic double bonds, as in $R_1R_2C{=}CR_3R_4$ and acetylenic triple bonds, as in $R_1{-}C{\equiv}C{-}R_2$, are examples. These particular features are also difficult to detect from hydrogen nuclear magnetic resonance spectra or ultra-violet spectra.

Pure rotational Raman spectra

As well as vibrational energy, molecules are capable of extracting rotational energy from the incident radiation and may indeed extract exclusively rotational energy, at least for gases where the rotational energy is quantized. Although the energy levels are the same as those which are relevant for the microwave spectra, the selection rules are different and so are the dominant observed frequencies. For linear molecules in the Raman effect the selection rule is $\Delta J = \pm 2$; the $+2$ refers to the Stokes line and the -2 to the anti-Stokes line, which may be of comparable strength in this case. The requirement of microwave spectroscopy that the molecule shall have a dipole moment is replaced by the requirement that the molecule shall have an anisotropic polarizability. This means that most molecules including nitrogen, N_2, acetylene, $H{-}C{\equiv}C{-}H$, and benzene, C_6H_6, can be studied, but not molecules which have a spherically symmetrical polarizability ellipsoid such as methane, CH_4.

Since the polarizability is independent of the sense of an electric field, a rotation of only π brings the polarizability ellipsoid into identity with itself in contrast to the rotation by 2π required for a dipole. In a classical form of radiation theory this would imply a Raman shift at twice the angular rotation frequency; this is analogous to the quantum selection rule $\Delta J = \pm 2$ rather than ± 1. For asymmetric molecules and for vibration–rotation lines, transitions with $\Delta J = \pm 2$, ± 1 or 0 may be allowed.

There is one point which arises with symmetrical molecules; it concerns the degeneracy or statistical weight of the rotational levels. For unsymmetrical molecules the nuclear spin degeneracies, $(2I+1)$ for each nucleus, and the rotational degeneracies, $(2J+1)$ for linear molecules, are multiplied together in the normal way. But if identical nuclei are symmetrically situated in the molecule, the nuclear statistics must be inspected. If the nuclei are of odd mass number, and hence of half-integer spin, Fermi–Dirac statistics apply and the total wave function must be antisymmetric with respect to the interchange of a pair of identical nuclei. For nuclei of even mass number, and hence of zero or integer spin, Bose–Einstein statistics apply and only symmetric total wave functions are allowed. Most ground electronic states and ground vibrational states are associated with symmetric electronic and vibrational functions, so that the important features lie in the rotational and spin factors.

For the particular case of homonuclear diatomic molecules, the even values of J relate to symmetric rotational functions and odd J to antisymmetric states. Of the $(2I+1)^2$ spin states, $(2I+1)(I+1)$ are symmetric and $(2I+1)I$ are antisymmetric. Consequently the ratio of J (odd)/J(even) spin degeneracies is $(I+1)/I$ for half-integer spins and $I/(I+1)$ for integer spins; a special case of the latter is $I=0$ for which all odd J states are absent. These spin factors are additional to the regular $(2J+1)$ degeneracies.

The best-known example is hydrogen where the ortho states with odd J have three times the degeneracy of the even J or para states. Interconversion is not brought about by any spectroscopic transition and in the absence of magnetic materials non-equilibrium mixtures of ortho and parahydrogen are stable for many months.

For nitrogen, $^{14}N_2$, and also for D_2, where $I=1$, the relative weights are 2:1 in favour of even J values. This alternation of intensity can be seen superposed on the Boltzmann distribution in the pure rotational Raman spectrum. Figure 9.3 shows the theoretical distribution for nitrogen. The individual line separations are $4B$, if centrifugal stretching is neglected, except for the central gap of $6B$ with the Rayleigh line in the centre.

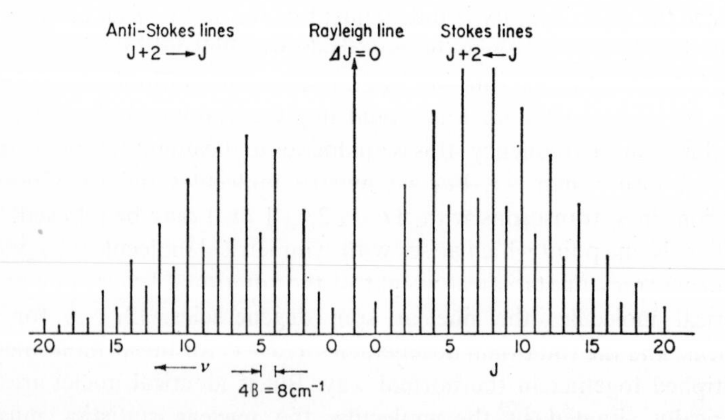

9.3 Theoretical intensities of a pure rotational Raman spectrum of N_2. (A photograph of an actual spectrum is reproduced in Herzberg, 'Spectra of Diatomic molecules', Van Nostrand, 1950, p. 64.)

The analysis of vibrational spectra

Chapters 8 and 9 have discussed the techniques and the simplest uses of infra-red and Raman spectroscopy. A complete understanding of vibrational spectra of any compound can yield considerable information about bond strengths, molecular shapes, the electronic structure, etc. The types of interpretation which lead to such information have been delayed until this chapter so that Raman and infra-red spectra can be considered together.

Diatomic molecules: energies

The simplest case is that of diatomic molecules. These have two atoms and only one fundamental vibration, namely that in which the two atoms move in and out together with respect to their common, stationary, centre of mass.

Classical dynamics gives the correct vibrational frequency in this and most other vibrational problems. The atoms are point masses m_1 and m_2 connected by the chemical bond which behaves like a mechanical spring with a Hooke's law elastic constant k. If x_1 and x_2 are the distances of the masses from the centre of gravity, their instantaneous separation $r = x_1 + x_2$; the equilibrium distance may be written r_e. Since the origin is the centre of gravity $m_1x_1 = m_2x_2$ and thence $x_1 = m_2r/(m_1 + m_2)$. The force exerted on the atoms by virtue of the bond distortion is $k(r - r_e)$ and this must be equal to the product of the mass and acceleration of atom 1. Then

$$k(r - r_e) = -m_1\ddot{x}_1 = -m_1m_2\ddot{r}/(m_1 + m_2) = -\overline{m}\ddot{r}$$

where $\overline{m}$ is written for the reduced mass, $m_1m_2/(m_1 + m_2)$. This is the equation of simple harmonic motion with a solution of the form

$$(r - r_e) = A \cos 2\pi\nu t,$$

with the frequency ν given by

$$2\pi\nu = (k/\overline{m})^{1/2}.$$

For gaseous beryllium oxide, BeO, $c^{-1}\nu = 1487\ \mathrm{cm}^{-1}$ and $k = 741\ \mathrm{kg\ s}^{-2}$, which are typical values for a light molecule with a fairly strong bond.

It is apparent that large force constants lead to high frequencies and that heavy masses give lower frequencies. Although with polyatomic molecules the detailed relationships are more complex, these general tendencies with force constants and masses still apply.

The quantum mechanical solution to the problem likewise gives $2\pi\nu = (k/\overline{m})^{1/2}$. But whereas in the classical treatment the amplitude, A, can have any value, the corresponding quantum mechanical quantity is restricted to discrete values. These are such that the energy of the vibration can be expressed as

$$E = (v + 1/2)h\nu,$$

where v, the vibrational quantum number, is restricted to positive integers or zero. The origin from which this energy is measured is the energy of the configuration in which $r = r_e$ and the atoms are stationary. But since the lowest level corresponds to $E = h\nu/2$ no molecule can exist without vibrational energy on this convention. The quantity $h\nu/2$ is called the *zero point energy* of the molecule.

Diatomic molecules are sufficiently simple that more exact treatments are not unduly difficult. The commonest is based on the Morse potential function. This expression for the potential energy V as a function of distance r is

$$V = D[1 - \exp - a(r - r_e)]^2$$

and the graph of this function is shown in Figure 10.1. In the above form the energy origin is the energy of the equilibrium configuration when $r = r_e$. The curve is sometimes referred to the asymptote as the energy origin when

$$V = D \exp\ -2a(r - r_e) - 2D \exp\ -a(r - r_e).$$

The difference between the asymptote and the equilibrium energy is D, the dissociation energy. This is sometimes written D_e to distinguish it from D_0, the dissociation energy referred to the ground vibrational state. An exact solution of the quantum mechanical problem for the potential gives energy levels

$$E_v = (v + 1/2)h\nu_e - (v + 1/2)^2 hx\nu_e$$

where x is the anharmonic coefficient. a is a parameter which expresses the width of the minimum. The following relationships hold between the parameters

$$\nu_e = (Da^2/2\pi^2\overline{m})^{1/2},$$
$$x\nu_e = ha^2/8\pi^2\overline{m},$$
$$x = h\nu_e/4D,$$
$$k = 2Da^2.$$

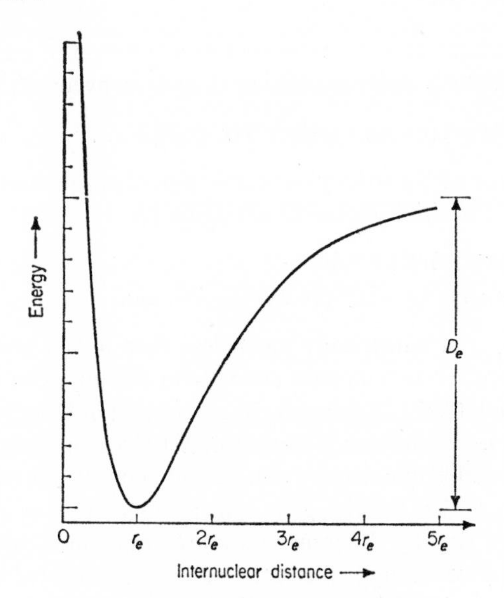

10.1 Morse potential energy function.

A knowledge of any two quantities, together with the reduced mass, enables the others to be derived. Usually the vibrational frequency is known from observation and the dissociation energy is inferred on thermochemical grounds or from the ultraviolet spectra. Alternatively x may be obtained by plotting successive vibrational intervals, ν_v, against the lower vibrational quantum number ν and obtaining the slope which is $2x\nu_e$ since

$$(E_{v+1} - E_v)/h = \nu_v = \nu_e - 2x\nu_e(v+1).$$

The potentials of actual molecules are moderately close to Morse curves, which are used for almost all approximate work. For the highest accuracy the energy levels are assumed to be a power series in $(v+1/2)$ with the next term $\nu_e y(v+1/2)^3$. Also there are discrepancies when $r \ll r_e$ as the Morse curve remains finite at $r=0$ whereas more exact potentials tend to infinity by virtue of the repulsion of the nuclear charges.

Diatomic molecules: intensities

The strength of infra-red absorption by different molecules varies quite widely and the factors governing intensity must be examined. As explained in chapter 2 the intensity of the $i \leftrightarrow j$ transition is governed by the transition moment

$$p_{ij} = \int \Psi_i p \Psi_j \, d\tau.$$

The dipole moment must be expressed as a function of internuclear distance and a Taylor's series expansion is most convenient, namely

$$p = p_e + (\partial p/\partial r)(r - r_e) + (\partial^2 p/\partial r^2)(r - r_e)^2/2 \ldots$$

For an oscillator with a strictly harmonic potential function the $(v+1) \leftrightarrow v$ and $(v+2) \leftrightarrow v$ transition moments are given by

$$p_{v+1,v} = [(v+1)h\nu/2k]^{1/2} \, \partial p/\partial r,$$
$$p_{v+2,v} = (v+1)^{1/2}(v+2)^{1/2}(h\nu/2k) \, \partial^2 p/\partial r^2, \text{ etc.}$$

Normally $p_{v+2,\,v}$ is numerically much less than $p_{v+0,\,v}$ and only transitions of the type $v+1 \leftarrow v$ appear strongly in the infra-red and these are each at the fundamental frequency. In absorption at laboratory temperatures the $v = 1 \leftarrow 0$ transition is the strongest since the great majority of molecules in equilibrium occupy the ground state. If there is appreciable anharmonicity $v = 2 \leftarrow 1$ and higher transitions may be resolved as their frequencies are slightly different; they are sometimes called *hot bands* since their relative strength increases at higher temperatures.

For a homonuclear diatomic molecule, as H_2, N_2, Cl_2, etc., $p = 0$ at all internuclear distances and all vibrational transitions have zero intensity in the infra-red. If there is resolved rotational fine structure, the individual line intensities are weighted by appropriate Boltzmann factors, but the total absorption is independent of the distribution between rotational lines. Hot bands are normally included in the experimental integration over the band and the integrated absorption over the fundamental band

$$\int \varepsilon d \ln \nu, \text{ is proportional to } p_{10}^2,$$

where ε is the extinction coefficient defined in chapter 8 on p. 103 and the integral is taken over a frequency range which includes all the absorption due to the fundamental and excludes all other bands.

There is no special difficulty in measuring absolute extinction coefficients and so $(\partial p/\partial r)^2$ can be measured. The sign of $\partial p/\partial r$ may be either positive—that is the same as p_e—or negative and the correct sign is not at all easily determined. Diatomic molecules in the ground state dissociate into neutral atoms and the dipole moment is expected to be zero at $r=\infty$ as well as at $r=0$ which corresponds to coalesced nuclei. Consequently the complete function must be somewhat as in Figure

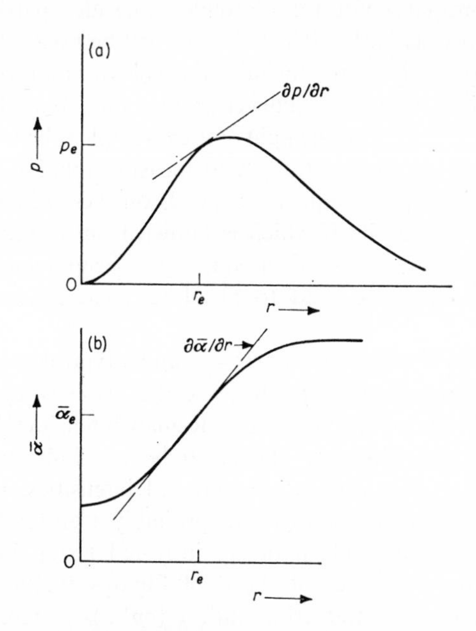

10.2 (a) Dipole moment and (b) electronic polarizability as a function of internuclear distance for a diatomic molecule.

10.2(a). The more polar the molecule the larger the slope $\partial p/\partial r$ is likely to be and the stronger the band. However, this rule is not universal since if the maximum of p is at r_e, $\partial p/\partial r=0$ at this point and the absorption intensity would be zero. Indeed $\partial p/\partial r$ is a quantity which is easily varied and in some cases the intensity is considerably changed by variation of solvent, weak complex formation, etc., so that measurements on gases are always to be preferred. Values of $\partial p/\partial r$ are normally about one-fifth of the charge on an electron.

In the Raman effect the intensity is controlled by the polarizability tensor, $|a|$, whose elements α_{xx}, α_{xy}, etc., may also be expanded in a Taylor's series as a function of r. There are two polarizability functions which are invariant with respect to molecular rotation, namely the mean polarizability $\bar{\alpha}$ and the anisotropy γ. For a general molecule

$$\alpha=(\alpha_{zz}+\alpha_{yy}+\alpha_{zz})/3$$
$$\gamma^2=(\alpha_{xx}-\alpha_{yy})^2+(\alpha_{yy}-\alpha_{zz})^2+(\alpha_{zz}-\alpha_{xx})^2$$

while for diatomic molecules $\alpha_{xx}=\alpha_{yy}$ by virtue of the cylindrical symmetry. Wholly polarized Raman lines have an intensity proportional to $(\partial\bar{\alpha}/\partial r)^2$ and depolarized lines to $(\partial\gamma/\partial r)^2$. These extreme cases may occur with symmetrical polyatomic molecules (with r replaced by a suitable symmetry coordinate) but for diatomic molecules partially polarized lines are obtained as both $\partial\bar{\alpha}/\partial r$ and $\partial\gamma/\partial r$ will be non-zero.

The proportionality factor includes the known vibrational frequency, the exciting light frequency, concentration, etc, but also some geometrical factors relating to the light source, sample tube and the spectrometer which are extremely difficult to measure reliably. The Rayleigh scattering of unchanged frequency depends on these same factors in the same way and on $|\alpha|$ itself which is known from refractive index and Kerr effect measurements. Consequently measurements of Raman/Rayleigh intensity ratios lead to absolute measurements of Raman intensities.

$|\alpha|$ is not zero either for $r=0$ or $r=\infty$ and a typical curve of $\bar{\alpha}$ against r is given in Figure 10.2(b). It is believed that $\partial\bar{\alpha}/\partial r$ is regularly positive and comparatively independent of r. Raman intensities have not been extensively studied, but they appear to be less influenced by factors external to the molecule than are infra-red intensities. Indeed Raman intensities of organic groupings are probably transferable between a wider range of compounds than are infra-red group intensities. This accords with the difference of shape of Figures 10.2(a) and (b) since small distortions and interactions affect $(\partial p/\partial r)_{r=r_e}$ more significantly than $(\partial\bar{\alpha}/\partial r)_{r=r_e}$. These curves also show that $(\partial^2 p/\partial r^2)_{r=r_e}$ may be large but that $(\partial^2\bar{\alpha}/\partial r^2)_{r=r_e}$ is likely to be small, and consequently $\Delta v=2$ transitions are more commonly strong in infra-red absorption than in the Raman effect.

Diatomic molecules: rotational fine structure

In the gas phase where rotational energy is quantized this too may vary during a vibrational transition. For infra-red absorption the selection rule $\Delta J=\pm 1$ applies. For the $J+1\leftarrow J$ transitions the increased rotational energy in the upper state is $2B(J+1)$ and a series of evenly spaced lines corresponding to different values of J are obtained as in Figure 10.3 which shows the spectrum of hydrogen chloride. The $J\leftarrow J+1$ transitions are at $-2B(J+1)$, that is on the low frequency side of the pure vibrational transition. There is a missing centre as no transition occurs without change of rotational energy. The frequency of the band centre corresponds to the pure vibrational quantum and is the figure quoted in discussions where rotation is irrelevant. Molecules which have non-zero orbital angular momentum, such as NO, may have a central line as the selection rule $\Delta J=0$ or ± 1 then applies. If there is no change of moment of inertia in the transition, $B'=B''$, and the central $\Delta J=0$ line, if present, is sharp and the wing lines exactly evenly spaced. If $B' \neq B''$ then even for $\Delta J=0$ transitions the frequency depends slightly on J so

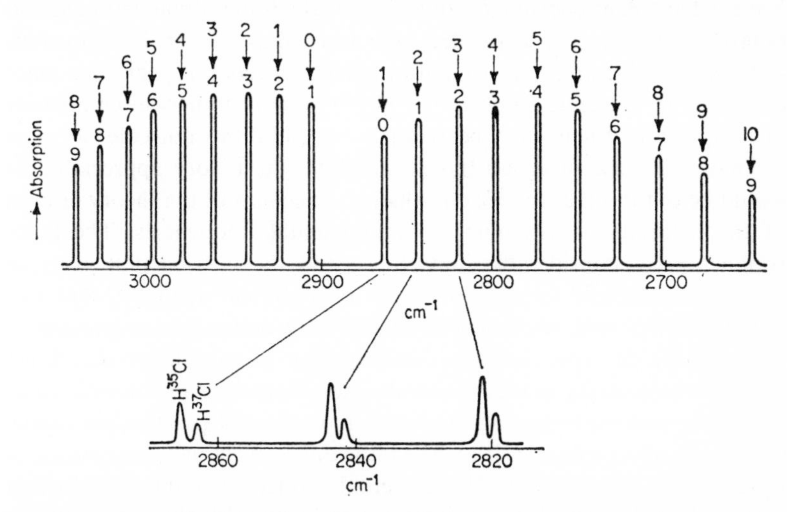

10.3 Rotational fine structure in the fundamental vibrational absorption in HCl gas. The lower part shows a section of the spectrum under higher resolution when the isotopic splitting appears.

that the central feature is broad or resolved into many lines; the flanking $\Delta J = \pm 1$ transitions are then unequally spaced (see also p. 113).

For the Raman effect $\Delta J=0$, ± 2 and the rotational lines are at 0, $\pm 2B(2J+3)$ with respect to the centre. The spacing is thus twice as large as in the infra-red.

The relative intensities of the individual lines are governed by the populations of the initial states. For absorption or Raman scattering the Boltzmann distribution applies providing degeneracies and factors related to nuclear spins are included as appropriate.

Sets of lines associated with particular ΔJ selection rules have been given special designations, namely $J'-J''=-2$, O-branch; $=-1$, P-branch; $=0$, Q-branch; $=+1$, R-branch; $=+2$, S-branch. These names apply in infra-red, Raman and optical spectroscopy.

Polyatomic molecules

For diatomic molecules there is only one degree of vibrational freedom and it is straightforward to pick out the fundamental spectrum. For polyatomic molecules there are commonly $3N-6$ degrees of freedom and as many fundamental frequencies and coordinates. To each fundamental vibrational mode there belongs a normal coordinate q which is a linear function of the Cartesian coordinates of the atoms; the detailed relationship can be determined if all the force constants are known. The q are independent coordinates in terms of which the dipole moment and polarizability may be expanded; the intensities of the fundamentals depend on $\partial p/\partial q$ and $\partial|\alpha|/\partial q$ and in many respects the q take the place of $(r-r_e)$ in diatomic spectra. The algebraic and numerical complexity of complete treatments may be considerable, but if the molecule has some symmetry the use of group theory simplifies the theory appreciably. It would be out of place to give an exhaustive account of the theory or even of the methods of assigning the fundamental frequencies. The latter problem is of considerable importance and the next section gives an illustrative example showing the types of argument available, and also the difficulties and uncertainties which may arise. An assignment is required for the spectroscopic determination of thermodynamic functions, for discussing related molecules, for discussing interatomic force constants, and for obtaining the dipole moment and polarizability functions from observed intensities. The immediate object in an assignment is to choose the complete set of fundamental frequencies and to confirm that every relevant spectral observation is consistent with the suggested set.

Vibrational assignment of cis-dichlorethylene

Figure 10.4 shows the structure of cis-dichlorethylene $C_2H_2Cl_2$. It is a

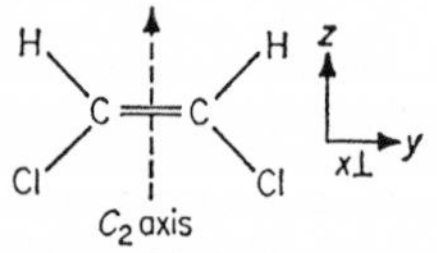

H H z
C=C
Cl Cl x y
C_2 axis

10.4 Coordinate system for cis-dichlorethylene.

flat molecule and the molecular plane is therefore a plane of symmetry which is, by convention, taken as the yz plane. The convention is of recent origin and some authors have x and y interchanged which results in an interchange of the b_1 and b_2 classes from the definitions used below. There is a second plane of symmetry, zx, which bisects the double bond and is perpendicular to it. These planes intersect in the z axis which is a two-fold axis of symmetry. Molecules which have just these three symmetry elements belong to the symmetry point group C_{2v}. For this point group the vibrations, and thence by association the frequencies and the normal coordinates, may be divided into four sets or classes designated a_1, a_2, b_1 and b_2. The vibrations differ as to which symmetry elements of the equilibrium molecule are retained throughout the vibration in the particular fashion indicated in Table 10.1. For a vibration of the a_1 class the full symmetry of the molecule is retained, for those of the b_2 class only the molecular plane yz is retained, for the vibrations of the b_1 class only the zx plane is preserved while for those of the a_2 class both symmetry planes are lost but the twofold z axis is retained. The vibrations of Figure 10.5 belong to one or other of these classes as indi-

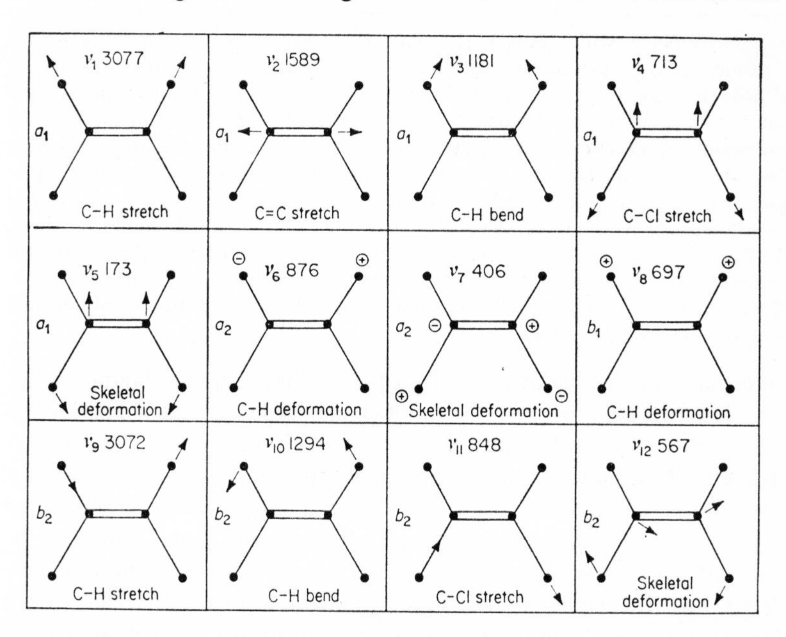

10.5 Qualitative descriptions of the fundamental vibrational modes of cis-dichlorethylene.

cated. If a symmetry plane is not retained the vibration must be antisymmetrical to the plane. This means that equivalent atoms related to each other move equal and opposite distances from the plane. Likewise for other symmetry elements 'antisymmetric' implies a loss of symmetry in an equivalent precise fashion in contrast to 'asymmetric' which means without symmetry; this is in accordance with the Greek roots anti- against and a- without as in antipathetic and apathetic. Table 10.1 also indicates the direction of the dipole moment change of the vibration, and hence its infra-red polarization, the type of polarization of the Raman lines and the classes occupied by the translational and rotational motions. These entries are pertinent for all molecules of C_{2v} symmetry and such

Table 10.1 *Symmetry information for point group C_{2v}; example cis-dichlorethylene*

CLASS	C_{2z}	σ_{zx}	σ_{yz}	p	T	R	RAMAN	NO.	SHAPE
a_1	s	s	s	p_z	T_z	—	pp	5	Doublet
a_2	s	as	as	ia	—	R_z	dp	2	ia
b_1	as	s	as	p_x	T_x	R_y	dp	1	Strong Q
b_2	as	as	s	p_y	T_y	R_x	dp	4	Weak Q
							Total	12	

C_{2z}, symmetry with respect to two-fold z axis, s = symmetric; as = antisymmetric; σ_{zx}, symmetry with respect to zx plane; p, activity of electric dipole moment transitions with direction indicated by the subscript, ia = inactive; T, class of translation of centre of mass in direction indicated by the subscript; R, class of rotation of the whole molecule about an axis indicated by the subscript; Raman, activity and polarization of Raman shifts, pp = partially polarized, dp = depolarized; No., number of vibrations in the class of cis-dichlorethylene; shape, the gas phase band contour for cis-dichlorethylene. For symmetry class of overtone or combination levels multiply classes of constituent vibrations according to the rules: $a \times a = b \times b = a$, $a \times b = b$, $1 \times 1 = 2 \times 2 = 1$, $1 \times 2 = 2$. Thus a state with one b_2 and one a_2 vibration singly excited belongs to the b_1 symmetry class.

information is readily available for other point groups in standard texts. The last two columns give the number of vibrations and indications of band contour and relate specifically to cis-dichlorethylene. The contours are related to the values of the principal moments of inertia. For such a heavy molecule most spectrometers are incapable of resolving individual rotational lines but the bands have a characteristic shape which can be calculated with some accuracy. For the geometry of cis-dichlorethylene $I_x > I_z > I_y$. In general if the dipole moment change is parallel to the largest moment of inertia there is a strong central Q-branch, if parallel to the smallest moment the Q-branch is weak and if parallel to the intermediate axis there is a central minimum.

Table 10.2 summarizes the experimental information on cis-dichlorethylene. There are five strong partially polarized Raman lines and the frequencies 173, 713, 1181, 1589 and 3077 cm^{-1} must be those of a_1 fundamental modes. Where possible fundamental frequencies refer to gas state observations, but it is so seldom that Raman frequencies relate to gases that average liquid state values are more commonly used and these will be quoted here. Infra-red band shapes only add confirmatory evidence for the a_1 fundamentals, but for the b_1 class there is only one strong band of even approximately the correct shape, namely at 697 cm^{-1}, although there is overlapping with the a_1 fundamental at 713 cm^{-1}. The very strong nature of the infra-red absorption makes it certain that this

Table 10.2 *Spectra of cis-dichlorethylene*

INFRA-RED			RAMAN			
VAPOUR	CLASS FROM CONTOUR	LIQUID	LIQUID	POLARIZATION	CLASS FROM POLARIZATION	ASSIGNMENT
		173	173 s	pp	a_1	a_1 ν_5
		–	406 s	dp		a_2 ν_7
571	b_2	571 s	563 m	dp		b_2 ν_{12}
697	b_1 ?	697 vs	–			b_1 ν_8
715	?	714 s	711 s	pp	a_1	a_1 ν_4
815	a_1	815 w	807 vw	?		$2\nu_7$
857	b_2	848 vs	–			b_2 ν_{11}
		–	876 w	dp		a_2 ν_6
		975 vw				$\nu_7+\nu_{12}$
		1022 vw				$\nu_5+\nu_{11}$
		1105 w				$\nu_7+\nu_8$
		1183 w	1179 s	pp	a_1	a_1 ν_3
		1218 w				$2\nu_8-\nu_5$
		1254 w				$\nu_7+\nu_{11}$
		1279 m				$\nu_4+\nu_{12}$
1303	b_2	1294 s	–			b_2 ν_{10}
		1404 w				$\nu_4+\nu_8$
		1470 w				$\nu_5+\nu_{10}$
		1571 w				$\nu_6+\nu_8$
1590	a_1	1590 s	1587 s	pp	a_1	a_1 ν_2
		1695 w	1689 w	?		$2\nu_{11}$
		1753 vw				$2\nu_6$
		2137 w				$\nu_{10}+\nu_{11}$
		2367 w				$2\nu_3$
		2753 w				$\nu_2+\nu_3$
2860	?	2855 w				$\nu_2+\nu_{10}$
3072	?	3072 s				b_2 ν_9
			3077 vs	pp	a_1	a_1 ν_1

Frequencies in cm^{-1}. w = weak, m = medium, s = strong, v = very, pp = partially polarized, dp = depolarized.

is a fundamental despite the absence of a Raman line which corresponds. A Raman line at 406 cm^{-1} is associated with transparency in the infra-red and must relate to an a_2 class fundamental; the other suggested frequency is 876 cm^{-1} which likewise is associated with infra-red transparency. Three of the strongest infra-red bands not yet assigned have b_2 contours and must represent fundamentals in this class at 567, 848 and 1294 cm^{-1}. The fourth b_2 vibration is essentially a C–H stretch and must lie just above 3000 cm^{-1} and the infra-red absorption at 3072 cm^{-1} is probably associated with this mode as well as the adjacent a_1 vibration. No lines at 848 or 1294 cm^{-1} are apparent in the Raman spectrum but it is seldom that all frequencies permitted in this spectrum are in fact observed.

Other weak infra-red bands can be explained as overtones and combination bands as indicated in Table 10.2. Exact equality between the observed and calculated sum frequencies is not to be expected because of experimental errors and anharmonicity effects. The frequency 1218 cm^{-1} is attributed to a *difference band* in which molecules initially in the state $v_5 = 1$, $v_8 = 0$ are excited to $v_5 = 0$, $v_8 = 2$, but it may be an impurity band. Only low frequencies may be excited in the initial states of such difference bands, since only low-lying energy states are sufficiently populated to give appreciable intensity. Further support for the assignment can be obtained from cis $C_2D_2Cl_2$ and cis C_2DHCl_2; assignments for these molecules can be made in conformity with the assignment of cis $C_2H_2Cl_2$. In particular the *product rules* are obeyed. These rules apply to isotopic molecules and relate the product of the frequencies of each class of one isotopic species to the product of the frequencies of the same class of a second isotopic species and to the molecular geometry and isotopic masses.

Although accurate descriptions of the vibrations require a knowledge of the force constants, approximate motions can be readily given by a mixture of guesswork and experience. Figure 10.5 shows such mode diagrams and also gives a corresponding verbal description. In some cases the type of motion and their frequencies can be carried over to related compounds and gradually the tables of characteristic frequencies are assembled. The diagrams of Figure 10.5 are highly simplified and small motions are not shown; as a consequence the diagrams may imply translation of the centre of mass or overall rotation both of which must be absent for the exact mode. Likewise hydrogen motions are not indicated when their contribution to both the potential and kinetic energy is small although the absolute amplitudes are not particularly small. Indeed such diagrams are normally a guide to memory and to understanding rather than realistic representations.

There are several reasons why it is worth making an investigation such as that just described. Not least is the assurance that the basic theory is sound when a complete, self-consistent and satisfactory explanation can be given for such a wide variety of observations. The complexity of the problem is oversimplified in the present account, which was written when the complete assignment was known and troubles due to impurities, lack of instrument range, etc., had been overcome. There are in fact ten published papers on the infra-red and Raman spectra of cis- and trans-dichlorethylene and it was 1954 before the last major feature of the trans spectrum was satisfactorily settled. Agreement of the observed frequencies with approximate estimates from the characteristic values discussed in chapter 8 gives confidence in the reasonableness of the concept of characteristic frequency. In particular the strong b_1 out-of-plane deformation at 697 cm^{-1} is welcome confirmation of the low value of this mode in other cis disubstituted ethylenes. Table 10.3 shows the figures of the

Table 10.3 *Out-of-plane C–H deformation frequencies of 1, 2 disubstituted ethylenes*

CHARACTERISTIC INFRA-RED FREQUENCY TABLES	DICHLORETHYLENE	
	INFRA-RED ACTIVE	INACTIVE
cis 675–728	697	876 cm^{-1}
trans 965–980	895	763 cm^{-1}

cis and trans cases. A set of valency force constants can be found which, when substituted into the appropriate formulae, lead to calculated frequencies in agreement with the observed. The constants are the same, within one or two per cent, for cis- and trans-dichlorethylenes. The difference between the frequencies of the very strong infra-red absorptions due to out-of-plane C–H deformations in the two cases arise wholly from the difference of molecular shape and, as can be seen, is counter to the Raman active, infra-red inactive, frequencies whose relative magnitudes are reversed.

Force constants

The force constants are also of interest as a necessary preliminary to the interpretation of intensities and to the drawing of more accurate normal mode diagrams than those of Figure 10.5. They have an interest in themselves as a measure of the strength of the chemical bonds. Unfortunately the relation of bond stretching force constants to bond properties is not straightforward, since a force constant is not strictly a bond property in a polyatomic molecule. Mathematically it is the partial second derivative of the molecular potential energy with respect to the bond length, $\partial^2 V/\partial r^2$, and, as in all partial differentiation, all other independent coordinates are to be kept constant. But the numerical value on differentiation with respect to one chosen coordinate depends on the nature of the remaining coordinates which may be chosen in more than one way for each molecule. The most significant scheme is that in which all other bond lengths and inter-bond angles are constant and under these conditions the force constant is effectively a bond property whose value may be compared in different molecules. Difficulties occur in cyclic compounds when there is a geometrical relationship between bond lengths and angles such that, as in benzene, it is not possible to stretch only one ring bond and keep all angles constant. Also forces between bulky, indirectly bonded atoms, such as the bromines in tetrabrom-methane, CBr_4, may be important and difficult to treat in a systematic way between different molecules. Despite these difficulties there is a fair correlation with other criteria for bond strength. Some typical values are given in Table 10.4.

Table 10.4 *Bond stretching force constants* (k/kg s^{-2})

Bond	k
⩾C–H	500
⩾C–C⩽	450
>C=C<	1000
–C≡C–	1600
>C=O	1200
⩾C–F	600
⩾C–Cl	350
⩾C–Br	300
⩾C–I	250

Thermodynamic properties

It is shown in textbooks which cover statistical thermodynamics, that such quantities as entropy, free energy, specific heat, etc., of pure gases can be related to partition functions (sums over states; *Zustandsumme*). These in turn are related to vibrational frequencies as well as moments of inertia and atomic masses. If all these are accurately known the calculated thermodynamic properties may be of greater precision than the directly measured calorimetric values, especially for high temperatures.

If a molecule has an internal rotational degree of freedom, this too makes a contribution, in which case a comparison of the observed quantities with those derived from vibrational frequencies leads to a measure of the height of the potential energy barrier opposing internal rotation. For many years this was the only way to estimate these barriers, but methods based on microwave spectroscopy are proving more accurate where they are applicable.

Free energies derived from spectroscopic data are the changes from absolute zero to the temperature in question and must be supplemented by heats of reaction if free energy changes and equilibrium constants are to be obtained. The enthalpy changes may be found from heats of combustion. Consequently equilibrium constants can be computed. This is of particular importance for equilibria which cannot be established in the laboratory.

For isotopic reactions, as for instance $DCN + HCl \rightleftharpoons DCl + HCN$, the enthalpy change would be zero if molecules were formed in their equilibrium configurations. For real molecules the zero point energy must be included; this is known from assigned frequencies so that for such reactions the equilibrium constant can be obtained wholly from spectroscopic measurements.

In the case of cis- and trans-dichlorethylenes the equilibrium constant has been measured. Using the assigned frequencies, the heat of isomerization can be deduced since all other quantities for obtaining the equilibrium constant are available from spectroscopic quantities. The value is found to be $1{\cdot}86 \pm 0{\cdot}01$ kJ mol^{-1} with the cis isomer more stable. The accuracy of this value is better than can normally be obtained by combustion or hydrogenation in a calorimeter.

Consequences of anharmonicity

The concept of a set of fundamental vibrational frequencies and associated normal motions with individual energies of the form $(v_i + 1/2)h\nu_i$ is central to most discussions of vibrational problems both in the general and the particular. However, the concept is only rigidly justified for a harmonic potential energy, that is when the potential, V, can be expressed in the form

$$V = \Sigma_i \, k_{ii} q_i^2/2 + \Sigma_{i>j} \, k_{ij} q_i q_j$$

where the q are Cartesian coordinates or linear functions thereof. Any terms of the forms $k_{iij} q_i^2 q_j$, $k_{ijk} q_i q_j q_k$, $k_{iijj} q_j^2 q_i^2$, etc., destroy the basic idea of fundamental modes both in classical and quantum mechanics. Such potential energy terms are said to be *anharmonic*.[1] For most molecules such terms are quite small and their influence, if not negligible, may be added at a late stage as a small correction calculated by perturbation methods. However, difficulty arises where two energy levels are, by chance, close together; this circumstance is sometimes called accidental degeneracy. Most important is the case $\nu_i + \nu_j \simeq \nu_k$ when small anharmonic terms may lead to a mixing of the $v_i = 1$, $v_j = 1$ combination level and the $v_k = 1$ fundamental level. Writing Ψ_{110} and Ψ_{001} for the vibrational wave functions respectively for these states and E_{110} and E_{001} for the energies then

$$\Psi_a = \Psi_{110} \cos\theta + \Psi_{001} \sin\theta,$$
$$\Psi_b = \Psi_{110} \sin\theta - \Psi_{001} \cos\theta,$$

[1] A few authors use the expression 'electrical anharmonicity' when referring to the higher terms, e.g. $\partial^2 p/\partial q_1 \partial q_2$, in the dipole moment expansion, but this nomenclature does not seem wholly appropriate.

where a and b subscripts refer to the mixed states and θ is a mixing parameter. $\theta = \pi/4$ expresses perfect mixing and θ approaches this value as $E_{110} - E_{001} \rightarrow 0$. The centre of the levels is unchanged as $E_a + E_b = E_{110} + E_{001}$ but the levels are somewhat further apart as $|E_a - E_b| > |E_{110} - E_{001}|$. The transition moments from the ground state to a and b are mixtures of the moments to the 110 and 001 levels but the total intensity is unchanged. In particular if the transition moment for the pure fundamental were large and for the combination transition zero, then the relative strengths of transitions to a and b tend to become equal as $\theta \rightarrow \pi/4$. Consequently two strong spectral features are found where only one was expected. The phrase intensity borrowing is sometimes used in this connection, but it must be remembered that the mixing is inherently in the energy levels and their wave functions and not really a mixing of transitions. In particular a difference band can never borrow intensity from an adjacent fundamental transition since there is no near equality of energy levels. Only levels of the same symmetry can be mixed in this way.

The whole process is sometimes called *Fermi resonance*; its occurrence is not infrequent, but it is almost impossible to predict just when it will occur. Even if the energy levels are nearly equal the requisite term in the potential energy may be too small to provide significant mixing. The possibility of Fermi resonance must always be kept in mind even in simple organic applications. Thus a compound having two $>C=O$ stretching bands may yet have only one $>C=O$ group. The converse case of only one infra-red band for compounds with two $>C=O$ groups may arise if there is such symmetry that one of the frequencies is allowed only in the Raman effect.

In the cis-dichlorethylene spectrum in Table 10.2, it is possible that the rather strong $\nu_4 + \nu_{12}$ combination band at 1279 cm^{-1} may have borrowed intensity from the ν_{10} fundamental at 1294 cm^{-1}.

Spectra of crystals and liquids

A further important method of assigning infra-red active frequencies to symmetry classes arises from the study of the absorption of single crystals using polarized radiation. For oriented molecules only radiation with an electric vector darallel to the direction of the change of the molecular dipole moment during the vibration is absorbed. If the crystal structure is known, and if all the molecules are aligned parallel to each other with a known orientation to the external crystal faces, the determination is straightforward. Most cases are more complex and molecules may occur in more than one orientation, but often the information may still be deduced.

In crystals too the interactions between molecules may be sufficient to distort the molecule and alter its symmetry. Under these circumstances frequencies which are forbidden for the free molecule may appear in the spectrum. Also in crystals there may be a coupling between molecules and a consequent splitting of the bands into two or more closely spaced features. In effect the whole unit cell is then the unit which must be considered in counting degrees of freedom and applying selection rules. Usually frequencies which are forbidden in the gas only appear very weakly in the solid.

Weak lines in violation of the selection rules may also appear in liquids. Since a typical spectral line may be 5 cm^{-1} in width, the absorption process may be said to occur in about 2×10^{-12} s. If a molecule is slightly distorted by intermolecular forces, the distortion may persist for this time in the liquid, since the patches of local order—or in this context disorder—may themselves exist this long. Consequently the selection rules for the distorted state apply to the molecule for this section of time. The expression *breakdown of selection rules in the liquid state* is used to refer to these apparent anomalies.

11 Atomic spectra

It is preferable to consider the electronic spectra of atoms before molecules. Although atomic spectra are the simpler because neither vibrational nor rotational energies need be considered, there is the complexity associated with spin and orbital angular momenta. Historically the observation of regularities in atomic spectra played an important part in the development of quantum theory. Now that the quantum theory is well developed, it is more logical to refer first to the energy levels and then to show what spectral regularities would be expected for transitions between them.

Hydrogen atom

Quantum theory predicts that a system of one nucleus of hydrogen and one electron has energy levels given by

$$E_n = -\mu_0^2 \overline{m} e^4 c^4 / 8h^2 n^2$$

where n is an integer known as the principal quantum number, e is the charge on the electron, and $\overline{m}$ is the reduced mass of the two particles, i.e. $\overline{m} = m_e m_p/(m_e + m_p)$. This may also be written $E_n = -hcR_H/n^2$ where $R_H = \mu_0^2 \overline{m} e^4 c^3/8h^3$ is the Rydberg constant for hydrogen and has the value $1{\cdot}096\,775\,81 \times 10^7\ m^{-1}$. This is one of the most accurately known functions of the fundamental physical constants and is used in their evaluation. Although very small departures from this energy formula are known and understood, it is obeyed with considerable accuracy. Quantum theory also shows that the energy levels are degenerate and that for a complete description of the eigenstate further quantum numbers must be introduced. These may be shown to be (i) the value of the orbital angular momentum, l. Strictly the square of the orbital angular momentum is $l(l+1)\hbar^2$ as with other angular momenta. l is a positive integer which may have any of the values 0, 1, 2, . . . $(n-1)$ for a level of principal quantum number n. (ii) The resolved orbital angular momentum, M_l, along a unique z direction. M_l is an integer which may have any value in the range l, $(l-1)$, $(l-2)$ . . . 1, 0, -1, . . . $-l$. (iii) The resolved electron spin angular momentum, M_s, along the z direction. Since $s=1/2$ for an electron, M_s is restricted to $+1/2$ or $-1/2$ for a one electron atom. (iv) The resolved nuclear spin angular momentum, M_I. For a hydrogen nucleus for which $I=1/2$, M_I is restricted to $+1/2$ and $-1/2$. Under some conditions these quantum numbers are not wholly appropriate since they no longer refer to strictly quantized degrees of freedom, but they do form a suitable set for most discussions. Besides the discrete negative energies corresponding to different values of n, all positive energies are possible. These positive solutions represent an unbound or ionized electron, which may have any amount of kinetic energy.

To deduce the spectra it is necessary to know the selection rules governing the transitions. Any change of n is permitted and since the energy is to a good approximation independent of the other quantum numbers, their selection rules do not affect the gross features of the spectrum. For electric dipole transitions these are $\Delta l=\pm 1$, $\Delta M_l=0$, ± 1, $\Delta M_s=0$, $\Delta M_I=0$. The transition frequencies are given by

$$\nu=(E'-E'')/h=R_{\mathrm{H}}(1/n''^2-1/n'^2).$$

This expression was determined empirically by early spectroscopists who had picked out series of lines with the same value of n''. For hydrogen the earliest known were the series with $n''=2$ which falls in the visible region; this series is known as the Balmer series after the scientist who, in 1885, formulated the wavelengths in essentially these mathematical terms. The corresponding series with $n''=1$ is the Lyman series which lies in the ultraviolet, while the series for $n''>2$ lie in the infra-red. In each of the series the spectral lines get closer and closer as $n'\to\infty$ and finally converge to a limit at R_{H}/n''^2 which is followed by an ionization continuum. The $n''=1$, or Lyman, series refers to a lower state which is the ground state and the corresponding energy of the edge of the continuum, R_{H}, is the ionization energy.

A correct description of the eigenstates is of importance both for examining the effect of magnetic and electric fields and for providing a large amount of information in a clear notation. The notation is also largely appropriate to atoms with more than one electron. The two angular momenta l and s, referring to the orbital motion and the spin, can be compounded to give a resultant angular momentum, j. With only one electron $s=1/2$ and $j=l\pm 1/2$. The resolved angular momentum along a unique direction, z, is M_j. $M=M_l+M_s$, the sum of the resolved orbital and spin momentum. If $l=0$, j must be 1/2 since negative angular momenta are not possible though the resolved momenta may have negative values. j may be further compounded with the nuclear spin momentum, I, which for H nuclei equals 1/2, to give the total angular momentum $F=j\pm 1/2$. Likewise the resolved momentum $M_F=M_j+M_I$. For the ground state of hydrogen $l=0$ so $j=1/2$ and there are two hyperfine levels with $F=0$ and 1 respectively. These levels differ in energy by 1420·405 751 78 MHz and this can be detected and very accurately measured in atomic beam experiments. Emission at this frequency is diagnostic for hydrogen atoms and is of great importance in radio-astronomy. It corresponds to a zero field electron resonance transition.

The abbreviated state nomenclature is given by quoting first the principal quantum number n, then the letter corresponding to the orbital angular momentum according to the scheme

$l=$	0,	1,	2,	3,	4,	5,	6,	7,	...
letter =	S,	P,	D,	F,	G,	H,	I,	K,	...

The value of the multiplicity, that is $2s+1$, is given as a left superscript and the value of j as a right subscript. F is not usually indicated. Thus the ground state of hydrogen would be written $1^2S_{1/2}$. The description of orbital angular momenta by letters is widely employed and should be memorized. After F for $l=3$ the letters follow serially in alphabetical order omitting J. When referring to single electron orbits, rather than atomic states, lower-case letters are used and one may talk of d electrons, meaning those in one electron orbitals having $l=2$. In many electron cases the one electron properties are indicated by the use of lower-case letters, n, l, j, etc., and the net, many electron, properties by the use of capitals, L, J, S, etc. One must be careful not to confuse S meaning a state with $L=0$, with S the value of the total electron spin.

Zeeman effect

One of the most powerful aids in analysing atomic lines is the application of a strong magnetic field. Such a field removes all the remaining degeneracies of the atomic energy levels and information is then obtained about the degeneracies of the levels involved in the transition. This splitting of the energies in a magnetic field, frequently called a Zeeman splitting, arises by virtue of the magnetic moment inherently associated with the orbital motion and the electron spin. The resolved orbital magnetic moment is $\mu_B M_l$, where μ_B is the Bohr magneton $=e\hbar/2m_e$. The energy from its interaction with a magnetic induction B in the z direction is $\mu_B M_l B$. For electrons the inherent magnetic moment is

$g_e\mu_B$ and the energy $g_e\mu_B M_s B$, where g_e is the free electron g-factor equal to 2·0023.

These energies only apply when M_s and M_l are good quantum numbers. When only M_j is appropriate, that is when neither l nor $s=0$, the energies are of the form $g\mu_B M_j B$; with the approximation $g_e=2$,

$$g=\frac{3j(j+1)-l(l+1)+s(s+1)}{2j(j+1)}.$$

Here g is known as the *Landé splitting factor.* For atoms j, l and s must be replaced by the total electronic quantities J, L and S.

The selection rule for allowed spectroscopic transitions is $\Delta M_J=0, \pm 1$. There is a difference in the polarization properties of these transitions in that $\Delta M_J=\pm 1$ transitions are caused by absorption of light polarized perpendicular to the magnetic field. $\Delta M_J=0$ absorption is caused by plane polarized light with an electric vector parallel to the field, so that its direction of travel is necessarily perpendicular to the field. For emission the light emitted bears the same relationship to the field as for absorption. A field of about a tesla may be required to produce splittings of a few cm^{-1}.

In very strong fields both the orbital and the spin magnetic moments may couple to the magnetic field more strongly than they couple to each other. M_L and M_S are again good quantum numbers. Such cases are known as a complete *Paschen–Back effect*; the intermediate stage, an incomplete Paschen–Back effect, is more common.

The observation of nuclear terms is possible with very high resolution.

Stark effect

There is a corresponding disturbance of the energy levels and splitting of the spectral lines by an electric field, which is known as a Stark effect. The details are different and in most atoms the effect is second order being proportional to $M_J^2E^2$; that is the energy shift is independent of the direction of the electric field E and the sign of M_J. However, the special degeneracy of energy levels of hydrogen atoms with different values of L, but the same n, leads to a modification and in this case a linear Stark effect is possible and the transitions are split into two with a separation proportional to E, not E^2.

Isotope effects

One effect of isotopic changes is to alter the value of the appropriate Rydberg constant which depends on the reduced mass. Thus lines of

D atoms can be resolved from those of H atoms. Also all effects depending on the nuclear spin, *I*, and the nuclear magnetic moment are grossly changed by isotopic substitution. There are also corrections for the different nuclear radii in the different isotopes and if s electrons are promoted there are shifts of the order of 10^{-3} cm^{-1} which arise from effects very closely related to the chemical isomeric shifts, as discussed in chapter 6 on Mössbauer transitions.

Hydrogen-like ions

Other one electron species such as He^{+}, Li^{++}, Be^{+++} give spectra which are governed by just the same considerations as hydrogen except that the nuclear charge is now Ze where Z is the atomic number. The energy of the level with principal quantum number n is now given by

$$E_n = -\mu_0^2 \overline{m} Z^2 e^4 c^4 / 8h^2 n^2$$

with $\overline{m}$ the new reduced mass.

Experimental

Atomic spectra can be excited in a number of ways, the most common being by an electric discharge in a low pressure source of the element in question. Metals may be used in rod form as electrodes for an electric arc when their spectra appear strongly in the arc light. Various high frequency electrodeless discharges may be used to avoid metal contamination in non-metallic elements or else electrodes of pure graphite impregnated with a solution of the element may prove useful. Narrowest lines come from cool sources when broadening arising from the Doppler effect is reduced. Emissive techniques are the most common, but absorption may be used, especially when it is important to identify transitions associated with the ground state. In some sources, especially in flames, self-absorption can be apparent. In this case the rather broad emission lines from the hot central region traverses the cooler surrounding gas where absorption only of the central region of the line occurs, since the Doppler broadening and Stark effect fields which broaden the emission are absent from the outer regions of the source. A number of atomic lines are observed in emission from the sun and stellar sources. Most sources are rich in the spectra of positively charged atoms and the relative strengths of the differently charged species depends considerably on excitation conditions.

Nearly any high resolution photographic spectrometer in the near infra-red, visible, ultra-violet or vacuum ultra-violet can be used as appropriate for the observation. High resolution gratings can be used for resolution of better than 1 cm^{-1}, but Fabry–Perot interferometers are used for even higher resolution, especially with almost monochromatic sources with small splittings due to nuclear moments, isotope effects or Stark or Zeeman fields. Although normally such interferometers are made with known dimensions so that the wavelengths can be determined from the instrument geometry, it is possible to reverse the process and using light of known wavelength, measure the distance between two planes with extreme accuracy. Indeed the precision of such experiments exceeds the precision with which marks on the standard metre bar can be observed. This means that the standard metre is unsuitable for distance measurements of the very highest precision and the international definition of the metre has been redefined in terms of the wavelength of an atomic line which can be measured with the very highest precision. The metre is now defined to be 1 650 763·73 times the vacuum wavelength of a specified orange line in the spectrum of the isotope of krypton of isotopic mass 86 in the absence of electric or magnetic fields.

Analysis

Since each element gives its own set of atomic spectral lines, a high resolution spectrum can be used to identify the individual elements present. The sensitivity varies very considerably from one element to another, but may be as little as a few parts per million of the original sample. Such spectroscopic methods are very widely used for detecting trace elements in steel and other metals, rocks, soils, etc. For such purposes the fact that the sample is destroyed by the discharge is no disadvantage especially as a few micrograms of sample may be enough.

Accurate quantitative work has proved more difficult because of the troublesome calibration of photographic plate densities and the irreproducibility of conditions in an electrical discharge. These can be overcome or circumvented and quantitative analysis is widely used. Some modern instruments for routine steel analysis and the like use a very high dispersion grating so that each element has a line which is not overlapped by those of any other element. These key lines are separately focused on to their own slit and photoelectric detectors each with its own amplifier and output meter so that a direct reading is obtained of the amount of each of some forty chemical elements. Or this information can be fed to magnetic tape and a computer used for automatic control of the process.

If high quantitative accuracy is required it may be useful to use aqueous solutions held in a porous graphite electrode to form the source, rather than solid specimens in the electrodes.

Another accurate and reproducible technique is to use an atomic absorption system. The sample in aqueous solutions is introduced via a fine spray, known as an atomizer, into a cool flame burning under well standardized conditions. The light source is then a stable, hollow cathode, emission source of the element. This is much hotter than the flame so that absorption occurs on passing through the latter. Since the source is liable to contain more than one spectral line, it is preferable to use a spectrometer in front of the photoelectric detector, but quite a modest instrument is sufficient as the resolution is essentially controlled by the source. Such a system may be fairly cheap but it has the disadvantage that a new source is required for each element.

Many electron atoms

When there are many electrons in an atom the simplicity of the hydrogen atom spectra is largely lost. Although there are important inter-electronic energy terms and although the electrons are indistinguishable, one may base the classification on a scheme which imagines the electrons fed individually into hydrogen-like orbitals. Each such one electron orbital is classified as 1s, 3p, 4f, etc., as in the states of a hydrogen atom. For the ground state of a many electron atom, the electrons are fed into the lowest energy levels. However, the *Pauli exclusion principle* must be respected and no two electrons may be placed in any one electron orbital. Any two electrons must occupy orbitals differing in at least one quantum number, that is differing in n, l, M_l or M_s. M_s may have the two possible values $\pm 1/2$ so that if electron spin quantum numbers are neglected, two electrons can occupy an orbital of given spatial character. Often the M_l quantum numbers are not shown separately so the common abbreviation for the shell filling in say the ground state of carbon with six electrons is written $(1s)^2(2s)^2(2p)^2$ and of argon with eighteen electrons as $(1s)^2(2s)^2(2p)^6(3s)^2(3p)^6$ where the superscripts indicate the number of electrons which have orbitals of the indicated value of n and l but differing in the values of M_l and M_s.

For the spectroscopic purposes one is interested in the energy levels and sophisticated features associated with electron repulsion are important. Consequently it is a poor approximation to the total energy of an atom to sum the one electron energies of its electrons in separate hydrogen-like orbitals. A simple, and reasonably effective, way to allow for inter-electron repulsion is to neglect all specific features and use a hydrogen-like energy with an average nuclear charge represented by $Z_{\text{effective}}$ in place of the true Z. For the inner electrons in the 1s shell the full nuclear charge is fairly realistic and $Z_{\text{eff}} \sim Z$. For the next, 2s, shell the electrons are predominantly further from the nucleus than the two 1s electrons and cannot easily distinguish these electrons from the nucleus. Consequently the effective charge is nearer $Z-2$ although the screening of the nucleus by the inner electrons is less than complete and a rather higher value of Z_{eff} is to be preferred. In this way a value of Z_{eff} for each electron shell can be suggested. These are useful for many crude purposes and can be improved by comparison with observed spectral transitions using the one electron energy formalism

$$E_n = -hcRZ_{eff}^2/n^2.$$

For the same principal quantum number, n, Z_{eff} is the larger the smaller the value of l. Thus the 4s orbitals are more stable than the 4p which are more stable than the 4d, etc., in many electron systems. Indeed this inequality is such that the 5s electrons have about the same energy as the 4d and are certainly of lower energy than the 4f. Such devices as the use of Z_{eff} are not to be trusted for accurate work, but the energy calculations get more complex the more exactly the repulsion terms are included. Even modern high speed computers working on self-consistent field programmes including relativity corrections are not easily able to match the experimental accuracy obtainable in the measurement of spectroscopic transitions. The chief difficulty is that the computed quantity is essentially the energy of the atom relative to separated electrons and nucleus and may be accurately known for two states, whereas the comparison is made between the computed difference of these two very large quantities and the accurate spectroscopic observations.

The spectra of monatomic ions are of essentially the same nature as those of neutral atoms. Indeed in isoelectronic series, e.g. Si, P^+, S^{++}, Cl^{+++}, the spectra are extremely similar and differ only in numerical detail. The spectra of ions are especially prominent in spark sources and are occasionally called spark spectra in contradistinction to the arc spectra of the neutral elements. When it is desired to distinguish the ionic source a Roman numeral may follow the atomic symbol. I refers to the neutral element, II to the monopositive ion, III to the dipositive ion and so on. Thus Ni XII refers to Ni^{11+}. Negative ions are much less common in discharge conditions.

Spectra due to changes in the valency shell

For spectra in the visible and near ultra-violet, the transitions involve predominantly changes in the outermost or valence shell electrons. These electrons lie outside complete electron shells whose electrons are not disturbed. These inner electrons are always paired so that the net spin angular momentum is zero. Also there is a theorem which states that such closed shells also have spherical symmetry and zero net orbital angular momentum. In spectroscopic notation the complete or closed shells have a ^1S configuration. It follows that these shells may be neglected in discussing state designations and transitions which concern only the outer electrons.

The remaining electrons have to be considered more carefully and the details depend somewhat on the particular atom. For the lighter elements the *L–S* or *Russell–Saunders coupling* scheme is the best simple approximation. To describe the atomic ground state, the electrons are allocated to different one electron hydrogen-like orbitals to obtain the lowest possible energy according to the criteria discussed above in connection with Z_{eff}. The shell filling scheme is called a *configuration*. Thus $(1s)^2(2s)^2(2p)^2$ is the ground configuration of neutral carbon with its six electrons. Because the one electron energies are the dominant energy features, the ground configuration is well separated from excited configurations.

However, the one electron energies do not depend on the resolved angular momenta, M_l and M_s, and so the simple energy requirement is insufficient to specify the manner of filling degenerate one electron orbitals. It remains therefore to distinguish the ground eigenstate from the many different states which may correspond to the ground configuration. The first guide is *Hund's rule* of maximum multiplicity which states that for the lowest energy the electrons are to be placed in different spatial orbitals so as to give the greatest possible electron spin angular momentum. That is S, the total electron spin, equals half the number of unpaired electrons. The basis of this rule lies in the antisymmetric property of the wave function with respect to the interchange of electrons which is the Pauli principle. This tends to correlate the motions of the electrons so that those with the same spin keep apart and have therefore a lower electrostatic repulsion than do electrons of opposed spins with the same one electron spatial distribution but no time correlation. Having satisfied this rule for lowest energy the electrons are assigned so as to obtain the highest value of the angular momentum L. For a given distribution the states with the maximum value of S, and thereafter of L, may lie a few thousand cm^{-1} below the excited states of lower values of S.

Thus for a silicon atom with fourteen electrons in its ground state the first ten electrons fill the levels of $n=1$ and 2 to give the closed shell structure $(1s)^2(2s)^2(2p)^6$. Energy requirements force the next two electrons into the 3s shell with opposed spins and the final two into 3p orbitals to give the configuration $(1s)^2(2s)^2(2p)^6(3s)^2(3p)^2$. The 3p levels being degenerate the electrons have parallel spins so that $S=1$ and a triplet state is formed. The orbital moment for a p electron is 1 but the electrons cannot both occupy the $M_l = +1$ orbital and retain parallel spins. Consequently they must fill the $M_l = +1$ and $M_l = 0$ spaces to give a maximum $M_L = \Sigma M_l = +1$; this arrangement forms a component of an $L = 1$ arrangement. In spectroscopic nomenclature therefore the lowest or ground state of silicon is the ^3P component of the $(1s)^2(2s)^2(2p)^6(3s)^2(3p)^2$ configuration.

Even these requirements leave uncertain the relative directions of L and S and thus the value of their vector sum J. The electrostatic terms are irrelevant here, but the magnetic coupling between the electron moment and the orbital moment becomes important. If the last shell is less than half full the lowest level is that with anti-parallel L and S and the lowest J and the *multiplet*—that is the set of levels differing only in the value of J—is said to be 'regular'. For more than half-filled shells, the highest value of J is favoured and the multiplet is '*inverted*'. If the shell is exactly half full as in $(p)^3$, $(d)^5$ or $(f)^7$ configurations, $L = 0$ and there is no multiplet structure. Thus for Si with $L = 1$, $S = 1$, J which runs from $L + S$ to $|L - S|$ may have the values 2, 1 or 0 and the energy states are $^3P_2$, $^3P_1$, $^3P_0$. The configuration at $(p)^2$ is less than half full, the term multiplet is regular and $^3P_0$ is the ground state. The energy intervals are related to the coupling between spin and orbital moments which is dominated by the one electron spin-own orbit coupling which is written in the form $\zeta \boldsymbol{l} \cdot \boldsymbol{s}$ where ζ is the spin orbit coupling parameter. ζ depends strongly on the nuclear charge and on n and is typically a few hundred cm^{-1}. Table 11.1 gives a few values for valence shell electrons. It should be remem-

Table 11.1 *One electron spin-own orbit coupling parameters (ζ) for some valence shell p orbitals*

2p	Li	Be	B	C	N	O	F	
	0·2	1	11	29	76	151	270	cm^{-1}
3p	Na	Mg	Al	Si	P	S	C	
	11	20	75	149	299	382	586	cm^{-1}
4p	K	Ca	Zn, Ga	Ge	As	Se	Br	
	38	105	386, 551	940	1550	1688	2460	cm^{-1}

bered that any attempt to describe many electron systems in terms of one electron parameters is necessarily based on an imprecise logical basis and can only give approximate results.

When Russell–Saunders coupling applies, the selection rules for electronically allowed transitions are $\Delta L = 0, \pm 1$, $\Delta S = 0$, $\Delta J = 0, \pm 1$ except that $J = 0$ to $J = 0$ transitions are forbidden. In this approximation transitions involve only one of the electrons and for this $\Delta l = \pm 1$. Weak transitions disobeying these selection rules, except those for J, are not at all uncommon and their existence implies that the Russell–Saunders scheme is not accurately applicable. One of the best-known atomic lines, the ultra-violet mercury line at 2537 Å, is designated $^3P_1 \leftarrow {}^1S_0$ and therefore violates the selection rule $\Delta S = 0$. Even when the coupling is appreciably disturbed the total number of states derived from any electron configuration is unaltered and there can remain a clear one to one correspondence between the exact states and the Russell–Saunders scheme. It may then prove convenient to use these labels to distinguish the states even though L, S, etc., are not exact quantum numbers. In the absence of nuclear spin, J will remain a good quantum number. This is an example of a more general rule that in any species, be it atom or molecule with or without nuclear moments, the total angular momentum is a good quantum number in the gas state when magnetic fields are absent.

The most useful alternative scheme, especially for heavy elements, is *jj* coupling in which each valency electron has its individual spin and orbital moments, s and l, coupled to give a one electron value of j and these may be coupled in various ways to give the separate possibilities for J.

There is one selection rule which applies for all schemes. The energy states may be divided into even and odd levels according as the total wave function is unchanged or reversed in sign respectively by inversion through the nucleus. These are sometimes designated by g (gerade in German = even) or u (ungerade or odd) written as subscripts: other texts use a small o superscript for odd states. For transitions occurring through electric dipole radiation only $g \leftrightarrow u$ transitions occur and this rule is strictly obeyed; it is sometimes called the *Laporte rule*. The inversion through the nucleus is equivalent to a coordinate transformation $x \rightarrow -x$, $y \rightarrow -y$ and $z \rightarrow -z$. Examination of the detailed wave functions show that s and d one electron functions are gerade functions while p and f are ungerade functions. In taking products the usual rule applies, that is $g \times g = u \times u = g$ and $g \times u = u \times g = u$, so that only configurations with and odd number of p and f electrons taken together can lead to a u-state. Thus the ground configuration of silicon, $(1s)^2(2s)^2(2p)^6(3s)^2(3p)^2$, forms g-states, and the excited configuration, $(1s)^2(2s)^2(2p)^6(3s)(3p)^3$, forms u-states. Electric dipole transitions between the states of one configuration are forbidden by this rule as they are necessarily $g \leftrightarrow g$ or $u \leftrightarrow u$ in character.

Sodium spectrum

The spectra of virtually all the elements in many ionic states have been assigned and tabulated, so that the methods used to assign complex spectra are of diminishing importance. However, a brief account of the rather simple spectrum of a sodium atom illustrates the ideas involved.

The first column in Table 11.2 lists the wavelength of some of the

Table 11.2 *Principal lines of the sodium spectrum*

(λ/Å)	($c^{-1}\nu$/cm^{-1})	ATTRIBUTION
22 084	4 526·9	4p↔4s
22 057	4 532·5	
11 404·2	8 766·4	4s↔3p
11 381·6	8 783·7	
8 194·82	12 199·5	3d↔3p
8 183·30	12 216·6	
6 160·73	16 227·4	5s↔3p
6 154·21	16 244·6	
5 895·93*	16 956·2	3p↔3s
5 889·96*	16 973·4	
3 302·94*	30 266·9	4p↔3s
3 302·34*	30 272·5	
2 853·03*	35 040·3	5p↔3s
2 852·93*	35 042·8	
2 680·48*	37 296·5	6p↔3s
2 680·38*	37 297·8	
2 593·99*	38 540·7	7p↔3s
2 543·93*	39 298·5	8p↔3s

Refractive index of air 1·000 276
* =lines absorbed by sodium vapour

strong emission lines of the neutral sodium atom. Since the aim is to derive an energy level scheme, the transition frequencies, $1/\lambda_{vac}$, are more useful and are given in column 2. From an emission spectrum it is not clear which transitions involve the ground state; introduction of a bulb of sodium vapour between the source and the spectrograph shows that the lines marked with a dagger to be absorbed and they must involve the ground state. This has the shell filling $(1s)^2(2s)^2(2p)^6(3s)$ and only one state $^2S_{1/2g}$ arises from this configuration. Only the 3s electron is likely to be disturbed and the lines showing reabsorption are likely to be due to the falling back of the electron from excited p levels, which are the only transitions allowed by the $\Delta l = \pm 1$ rule. These lines occur largely as closely spaced pairs, but so do the energy states since J may be 3/2 or 1/2. These $^2P_{1/2}$ and $^2P_{3/2}$ states are separated by only a few cm^{-1} since the spin orbit coupling is small for sodium. It rapidly decreases for the higher states and by the time the lines involving 7p and 8p one electron orbitals are reached the splitting of the lines is below the spectrometer resolution. This enables a tentative principal quantum number designation to be put on the upper states and a table of energies as in Table 11.3 can be started.

If the lowest continuum can be observed or the energy levels otherwise related to the ionization limit a further check is possible. For outer levels

Table 11.3 *Energy levels for sodium*

POSITION OF LAST ELECTRON	$c^{-1}\nu/\text{cm}^{-1}$	STATE	Z_{eff}/n	n^*	Z_{eff}
$\infty = \text{Na}+$	41 449·6	$^1S_0$			
8p	39 298·5	^2P_u	0·1400	7·143	1·120
7p	38 540·7	^2P_u	0·1628	6·142	1·140
6p	37 297·8	$^2P_{3/2\ u}$	0·1945	5·142	1·167
	37 296·5	$^2P_{1/2\ u}$			
5p	35 042·8	$^2P_{3/2\ u}$	0·2417	4·137	1·208
	35 040·3	$^2P_{1/2\ u}$			
5s	33 200·7	$^2S_{1/2\ g}$	0·2741	3·649	1·370
4p	30 272·5	$^2P_{3/2\ u}$	0·3190	3·135	1·272
	30 266·9	$^2P_{1/2\ u}$			
3d	29 172·9	$^2D_{3/2\ g}$	0·3346	2·990	1·004
	29 172·8	$^2D_{5/2\ g}$			
4s	25 739·9	$^2S_{1/2\ g}$	0·3785	2·642	1·514
3p	16 973·4	$^2P_{3/2\ u}$	0·4722	2·119	1·417
	16 956·2	$^2P_{1/2\ u}$			
3s	0	$^2S_{1/2\ g}$	0·6146	1·627	1·849

Z_{eff} will be a little greater than unity as the central charge of 11 units is well screened by the 10 inner electrons. If one assumes $Z_{eff} = 1$ and hydrogen like orbitals one can obtain effective values for the principal quantum numbers, n^*, which must be lower than the true n since too low a value has been taken for Z_{eff}. However, these values of n^* are sufficiently regular to be useful in assigning n and in particular they vary smoothly (they are in fact 2·119, 3·135, 4·137, 5·142, 6·142, 7·143 for the p levels) which shows that no low p level is missing because its transitions escaped observation. Other lines involving these same ^2P levels can be recognized as they occur in close pairs with exactly the same splitting. These allow further states to be located and the $\Delta l = \pm 1$ rule requires that these have the electron in s or d states. In principle the ^2D levels could be distinguished by the spin orbit coupling which separates the $^2D_{5/2}$ and $^2D_{3/2}$ states, but the coupling in d shells is less than in p shells and the lines are not resolved by the spectrometer. However, the shielding is usually especially effective for d orbitals which scarcely penetrate the inner electrons and the level with $n^* = 2{\cdot}990$ can be assigned to the level with 3d electron and therefore to the lowest ^{2}D state on these grounds.

Figure 11.1 shows a graphical presentation of most of the information in Tables 11.2 and 11.3 in a form frequently used.

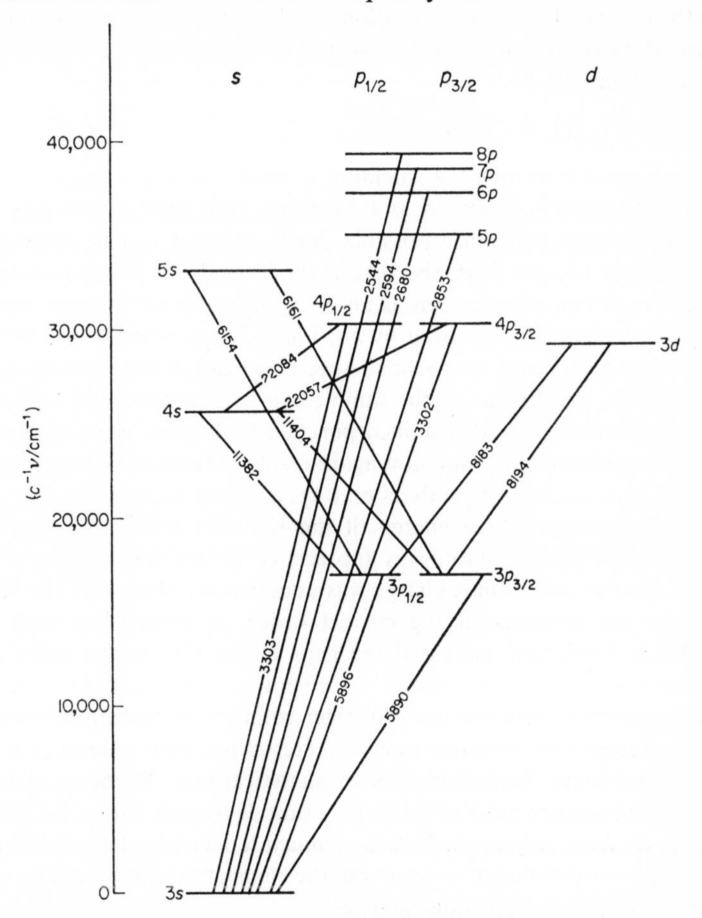

11.1 Diagram showing the one electron excitation state energies of the sodium atom and the wavelength in Angstroms of the allowed transitions.

Spectra due to inner electrons

If the inner electrons are disturbed the possibilities are more numerous, more difficult to disentangle and often more trouble to observe since they are liable to occur in the so-called *vacuum ultra-violet* (above 55 000 cm^{-1}) where air absorbs very strongly. Matters become simpler again if the transition arises wholly in the deepest electrons, since the influence of the outer electrons is then relatively small. The best-known transitions of this type are the Kα emissions resulting from the 2p$\rightarrow$1s one electron process. A strong doublet can be seen in the X-ray region; the position of its centre depends predominantly on the atomic number Z, but only slightly on the degree of ionization of the outer electrons or on the nature of the compound in which the atom may be bound. The frequency is given essentially by

$$c^{-1}\nu = Z_{\text{eff}}^2 R(1/1^2 - 1/2^2).$$

For such inner electrons the shielding is small and $Z_{\text{eff}} \sim Z_{\text{true}} - 1$. The doublet structure is observed as transitions from each of the $p_{1/2}$ and $p_{3/2}$ one electron states are possible. Such emission can be studied as typical X-ray sources with the tube cathode made of the test element. The emission can also be seen as an X-ray fluorescence process, one of the 1s electrons being removed by a primary X-ray beam. Fast electron beams may be focused on to small specimens and in specially designed sources quite small areas of the sample can be examined and analysed. Also some unstable nuclei can incorporate a 1s electron, a process called K shell capture and Kα emission again results. Standard X-ray spectrometers which use ionic crystals as gratings are used for the observations.

The dependence of the energy of the Kα lines with position of an element in the periodic table was appreciated before the meaning of the atomic number and atomic charge was understood; Mosely in 1913 thus classified the elements in the correct order, in conformity with the Mendeleev table, and indicated the gaps in the rare earths series and elsewhere, which have been filled by elements subsequently identified.

In recent years with the use of Geiger counters, scintillation counters and other improved detection systems, interest has been revived in X-ray spectra and X-ray fluorescence as an analytical tool. Elements of large atomic numbers are most suitable, for exam le barium with a Kα line at 0·387 Å (*c.* $2{\cdot}6 \times 10^8$ cm^{-1}). X-rays of much longer wavelength, the soft X-rays, are more difficult to detect but the elements as far as sodium with Kα at 11·9 Å have been thus analysed.

Ionization

The onset of a continuous absorption spectrum at the high frequency side of a set of lines indicates that an electron has been completely removed from the atom. The onset of the continuum corresponds to the ionization energy of the atom and beyond the edge the extra energy absorbed appears as the unquantized kinetic energy of the electron. Large kinetic energies are unlikely to occur and the absorption gets progressively weaker at higher frequencies. Further sharp spectral absorption lines may appear embedded to the continuum or at even higher frequencies. The related transitions occur from the excitation of electrons other than the most weakly bound one or, occasionally, to the excitation of two electrons simultaneously. At even higher frequencies further continua appear, whose edges correspond to the ionization limit when at least one electron remains in an excited state. The separation of these continua therefore correspond to the excited levels of the ions and can be checked from the ionic spectra.

Transition metal ions

In general the interaction of surrounding media with atoms in the liquid or solid states is so strong that few measurements are very informative. The same is true for many monatomic ions. However, the d electrons of transition metal ions and the f electrons of rare earth ions are well shielded and fairly sharp transitions are possible. These transitions inside the d and f shells are mostly of low energy and are related to the visible colours of most transition metal compounds; some transitions lie well into the infra-red. They are mostly weak transitions compared to allowed free atom transitions since they have $\Delta l=0$. However, the l, the orbital angular momenta, are not good quantum numbers in the absence of spherical symmetry and the g, u classification may also be lost. The electric fields due to other atoms in the neighbourhood of the ion have the greatest effect and the transition frequencies give information about the tightness of the binding of neighbouring ligands and the symmetry of their arrangement.

The details of these spectra may get quite involved, especially for systems of low symmetry, but the application of group theoretical methods have aided understanding and a fairly complete, consistent explanation of both the electronic absorption and the associated magnetic properties can usually be given. The theory is essentially an extension of that given for atoms when spherical symmetry is no longer assumed.

Lasers

The successful attainment of maser oscillations as described in chapter 7 led to work on optical transitions in the hope of obtaining similar devices emiting visible light. These use the same basic principle which requires that more atoms are in the upper state of a transition than in the lower so that induced emission is stronger than absorption. The name *laser* (Light Amplification by Stimulated Emission of Radiation) or optical maser is given to such devices. Although essentially amplifiers in the sense that more light is emitted than is required to induce the emission, like all amplifiers they can be converted to oscillators by introducing feed-back from output to input with the correct phase relationship. This feature is readily incorporated through a mirror system and laser oscillators are much commoner than the true amplifiers. Since the induced emission is in phase with the inducing light the strong coherent emission which builds up is extremely monochromatic and capable of forming a thin, intense, parallel ray.

The chief difficulty is the production of a sufficient number of excited state atoms. The first successful experiments by Maiman in 1960 achieved this by optical pumping using a large flash discharge which was wrapped round the central laser rod. This was the working material comprising chromium ions contained in a ruby crystal and the red, emitted light was at 6943 Å. In such devices the discharge, and hence the oscillations, are pulsed and last only a few milliseconds or less; repeat pulses cannot be too frequent as the working crystal gets hot and must be allowed to lose its heat to its surroundings by thermal conduction. The light during the pulse is intense, reaching many megawatts per square centimetre and is finding industrial use for drilling small holes. It is also used to vaporize small specimens for spectroscopic sources. The most intense pulses of all are obtained in '*giant pulse*' or '*Q-spoilt*' lasers in which the population in the excited state is allowed to rise before oscillation can occur. The onset of oscillation may be delayed by using a revolving mirror which is only in its correct position towards the end of the discharge flash causing the excitation.

It should perhaps be made clear that the energy of such a laser is derived from the initial discharge, whose energy per pulse is much greater than that of the laser pulse, but being distributed over a much greater solid angle and volume its radiation density in watts/square centimetre is less. This input or 'pump' light is quite distinct from any inducing light which would form the signal if the laser were used as an amplifier; such a signal would need to be of the same frequency as and colinear with the final laser beam.

The electric field in the light can be so strong that the polarizability of solid materials behaves in a non-linear fashion and frequency doubling can occur so that light of half the wavelength is produced.

For many purposes peak power is of less importance than constancy of frequency and continuous operation and then a gas laser is more valuable. The first such device was the helium-neon infra-red laser of Javan in 1960. A diagram of a more recent version is given in Figure 11.2. The gas-filled glass tube is placed between the reflecting mirrors through which the useful light escapes from the optical cavity formed by the mirrors. The tube windows are set at the Brewster angle, so that one

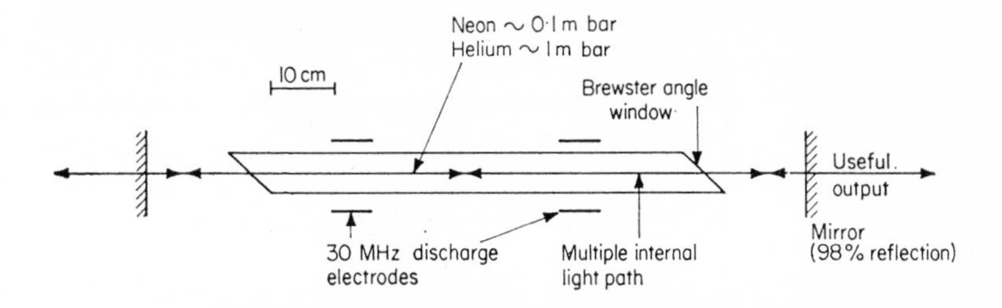

1.2 Simple diagram of a helium-neon continuous gas laser.

plane of light polarization is transmitted with great efficiency and the emitted light is highly polarized in this plane. An electrical discharge is maintained in the tube by means of external radiofrequency electrodes fed with about 100 W at 30 MHz. At higher powers stronger emission is observed but this is liable to contain several frequencies, and other light polarizations, and may be less valuable than threshold emission. Though the light frequency is essentially determined by the atomic gas, the exact value is controlled by the exact distances between the mirrors which must be an integral number of half wavelengths. The monochromatic nature of the light is controlled by the constancy of this distance and with ideal conditions the frequency may be constant to a few kHz in a frequency of 3×10^{14} Hz, that is to 1 part in 10^{11}. The actual inherent line widths may be even less.

The relevant energy diagram is shown in Figure 11.3. The lower excited states of helium correspond to the configuration (1s)(2s) and the ^{3}S component lies at 159 850 cm^{-1}. This corresponds to 20 eV and excitation to this level is efficiently brought about by the electric discharge. Since it is a triplet level of a light atom, radiative transitions to the singlet ground state are strongly forbidden and the ^{3}S state is consequently fairly long lived. Neon has an excited configuration, $(1s)^2(2s)^2(2p)^5(4s)$ at just about this energy. Collisions between ground state neon and excited helium are effective in causing radiationless transfer of energy to form ground state helium and excited neon. The higher states of neon are not accurately described by Russell–Saunders coupling, but there are four states of the configuration which span about 100 cm^{-1}. One in particular lies at 159 537 cm^{-1}; this is 313 cm^{-1} below the helium level and the radiationless formation of this state leaves only this 313 cm^{-1} of energy to be released into kinetic energy. A lower configuration of neon, $(1s)^2(2s)^2(2p)^5(3p)$, is not directly populated from the helium as

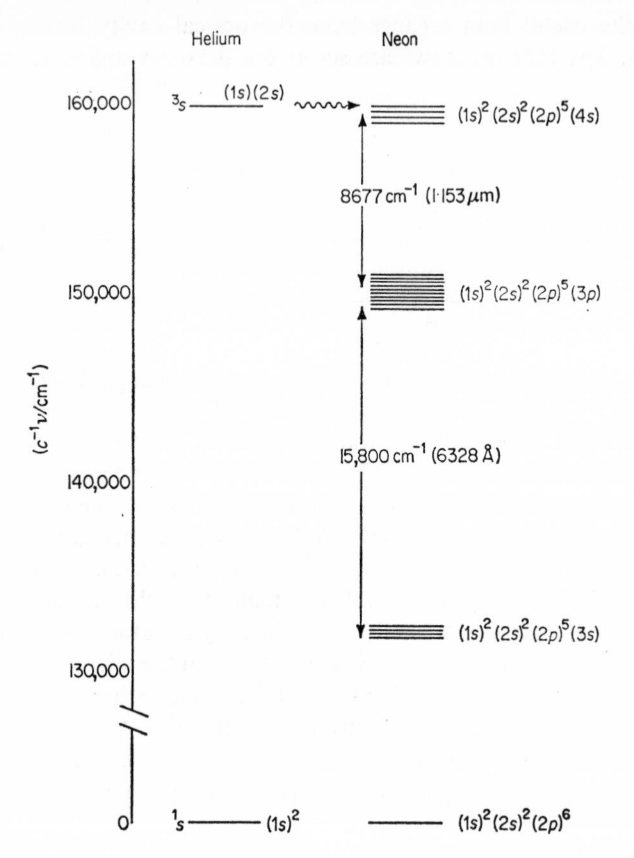

11.3 Excited states of helium and neon involved in a gas laser.

the energy mismatch of 10 000 cm^{-1} is too great to be readily transformed to kinetic energy. But radiative transitions between these two excited neon configurations are allowed and in particular transitions to a state at 150 860 cm^{-1}. The laser condition that the higher state be more populated is fulfilled and indeed the transition between the two specified levels with a frequency of 8677 cm^{-1} is the laser line observed at threshold power levels. Other laser lines between these configurations are observed at higher powers.

The reflecting mirrors need to be of high efficiency and special dielectric coatings whose reflectance is a maximum at this frequency. If these are replaced by mirrors suitable for reflection of red light, it is possible to get laser action at 6328 Å due to transitions between the $(1s)^2(2s)^2(2p)^5(3p)$ and the $(1s)^2(2s)^2(2p)^5(3s)$ configurations. Such lasers are finding many uses as light sources especially in metrology where accurate distance measurements are required. They are also useful sources for standard measurements of refractive indices, optical rotation, etc., and the wavelength of 6328 Å is beginning to replace the sodium D lines for such purposes. Laser sources are also useful in Raman spectroscopy and other experiments which depend on the scattering of light.

12 *Low resolution electronic spectra*

Electronic spectroscopy of molecules is a field of study which covers all electronic transitions whether the radiation involved is actually in the near infra-red, visible or ultra-violet region. This subject has attracted two rather disparate groups of practitioners. On the one hand there are the devoted spectroscopists whose aim is ever higher resolution and ever greater detail in the interpretational side and who concern themselves predominantly, though not exclusively, with diatomic molecules in the gas phase. Such studies are considered in detail in chapter 13. The other group are concerned with quantitative analysis, identification of large organic molecules and identification of key chemical groupings in unknown structures. This field has been of great benefit to organic chemistry, especially before the analogous, but more informative, studies of infra-red spectra and high resolution nuclear magnetic resonance were developed. Speed, convenience and small samples become important and low resolution spectra in solution are favoured. This present chapter will concentrate on these aspects, but it must be appreciated that there is no strict division.

Experimental

Many ultra-violet spectrometers bear a close resemblance to infra-red spectrometers and it is chiefly important to recognize the differences. Figure 8.1 on page 93 would represent the simplest form. For routine absorption measurements the sources are normally a tungsten filament lamp from 12 000 to 35 000 cm^{-1} (8000 to 3000 Å) and a high pressure hydrogen discharge lamp from 30 000 cm^{-1} to above 55 000 cm^{-1}. These are approximately the visible and near ultra-violet ranges. The lower end of the range represents the end of the interest for most compounds and special cases would be studied on an infra-red instrument. If the optical path is filled with air, the oxygen absorption band limits the upper frequency to 50 000 cm^{-1} (2000 Å), but in a closed system flushed with nitrogen operation to 55 000 cm^{-1} is quite feasible. This extra range is especially valuable for organic molecules with isolated double bands which absorb at about 52 000 cm^{-1}. Different glasses start to absorb in different parts of the near ultra-violet and so front surfaced mirrors are used for focusing. The shorter wavelengths demand a higher quality of reflectivity to avoid scattered light and the aluminium may be coated with a thin layer of magnesium fluoride to reduce deterioration in a laboratory atmosphere. Windows and cells are made of quartz or fused silicon and cells are typically 1 cm thick for solution work. Cells of accurately known thickness are readily made for quantitative work and they can be fitted with stoppers or with taps and side arms for attachment to a vacuum line.

For recording work the detector is usually some photoelectric device, most often a photomultiplier. In this the light falls directly on an active photosensitive surface such as caesium and each photon releases an electron. An electric potential accelerates this towards a second active surface where its impact may release several secondary electrons. These in turn are accelerated electrically to a third surface where again each electron releases four or five new electrons. The process is repeated for 10 to 15 stages each with 70–100 V accelerating potential and the total current amplification can reach several million. The current in the final stage activates the recording devices. The dark current, that is the current with no incident light, can be made very small, as can the associated noise. As a result more strongly absorbing solutions can be measured in the ultra-violet than in the infra-red. The noise is, for moderate light intensities, proportional to the square root of the light intensity and is due to the random instants of arrival of the individual light photons at the collector surface. The other important detector in this region is the photographic plate. This too has a favourable signal-to-noise ratio as it integrates the light intensity over the time of the exposure. The band width is, in effect, the reciprocal of the exposure time. It is thus extremely efficient for weak emission signals and has the advantage that a wide range of wavelengths can be observed simultaneously. However, the convenience and the accuracy of photoelectric detectors make them favoured for general low resolution applications.

As with infra-red spectrometers, modern commercial instruments reach a considerable degree of sophistication. Figure 12.1 shows a simplified diagram of a double beam recording instrument. Whereas in the infra-red the rough quality of solid samples favours a sample compartment before the front slit of the monochromatic section, it is often undesirable for a sample to receive the full strength of the ultra-violet source which may cause decomposition. With a sample behind the back slit only a small fraction of the source light reaches it at any instant, since all light of frequency other than that for which the instrument is set is stopped by the jaws of the back slit. For double beam operation the light is split into two beams one of which traverses the solution, S,

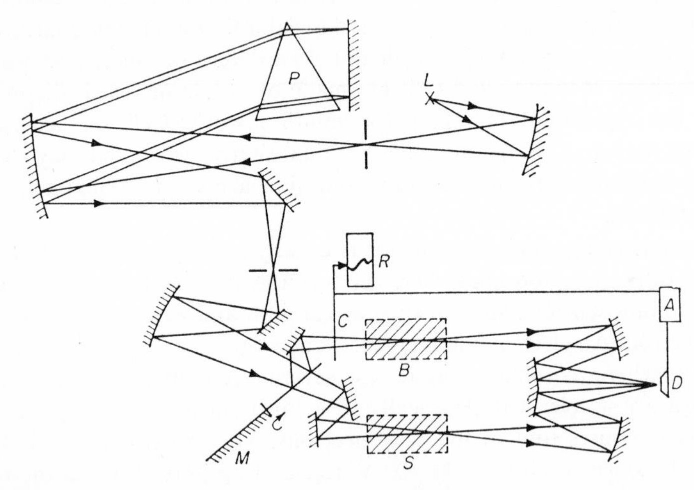

12.1 Diagram of a recording ultra-violet spectrometer. L, light source; P, prism; M, rotating semicircular mirror; C, comb attenuator; R, chart recorder; B, pure solvent cell in 'background' beam; S, solution cell in 'sample' beam; D, detector; A, amplifying circuits.

and the other the blank cell, B, containing pure solvent. This division can be made on a time basis by the chopper mirror, M. This alternately sends the light through path B and then via apertures in the mirror blade, through path S. The signals are recombined on one detector, D, which has therefore a signal fluctuating at the chopper frequency. A phase-sensitive detector in the amplifier senses which of the paths is the more transmitting and feeds a signal which removes or inserts the comb-shaped diaphragm, C, to maintain equality of transmission in the two paths. The extent of insertion of this comb indicates the absorption which is recorded on the chart record, R. The frequency is slowly, but continuously, varied by rotating the prism and its reflecting mirror and the chart is moved in synchronism, often being on a drum which rotates. If a suitable logarithmic comb or its equivalent can be made, the vertical axis reads optical density rather than percentage absorption. As with the infra-red instruments, automatic slit opening, prism rotation, automatic repetition, variable speed and time constant features can be incorporated. Some of these may have control settings, but the inescapable operations are merely loading the sample cell, inserting a chart paper and turning on the drive. The spectrum is then drawn on the precalibrated paper ready for interpretation.

A wide range of transparent solvents are available for the range up to 55 000 cm^{-1}. Water, heptane, chloroform and benzene-free alcohol are the most useful.

There is not a single uniformly adopted method of plotting the results. The most common practice is to plot the wavelength linearly increasing to the right and molecular extinction coefficient, ε, vertically. A linear frequency scale is certainly preferable; it should increase from right to left. Either cm^{-1} or sometimes electron volts are used. If a wide range of ε is encountered and it is desired to show the weak and strong features together, $\log_{10} \varepsilon$ may be plotted vertically.

The principles of analysis given under the infra-red chapter on page 102 hold for the ultra-violet also. Because of its convenience the visible and ultra-violet spectra are used to analyse a wide variety of products in the chemical industry, in the pharmaceutical industry and in research laboratories. The practice is also growing of turning inorganic materials into intensely absorbing complexes, often visibly coloured, and completing inorganic quantitative analysis spectrographically in solution.

Electronic transitions in organic molecules

The sensitivity of electronic absorption methods is high and formally forbidden transitions are often easily observed. As a result selection rules are of little guide to the practical usefulness of absorption spectra and one must rely largely on the expected order of magnitude of excitation energies. Saturated organic molecules seldom have transitions in the regular range below 50 000 cm^{-1}, but for unsaturated molecules there are low-lying excited states which can be reached by absorption in this region. The existence of a visible or ultra-violet absorption spectrum can be taken as evidence of unsaturation. For particular types of unsaturation the absorption regions are more closely defined and Table 12.1 gives a guide to typical frequencies for some common structures. It can be seen that there is considerable variation both in intensity and position.

Much is known about the effect of substituents on these bands. An increase of conjugation—that is alternate single and double bonds—always lowers and intensifies the main absorption bands as seen in Table 12.2 for the simple conjugated acids.

Substituents without further unsaturation have much less effect, though the changes which occur are regular. Some groups, including

Table 12.1 *Some characteristic electronic absorption frequencies*

TYPE OF COMPOUND	POSITION cm^{-1}	Å	PEAK INTENSITY ε
Acetylenes	58 000	1730	5×10^3
Ethylenes	52 000	1930	10^4
Benzenes	50 000	2000	8×10^3
	38 500	2650	$2 \times 10^2 - 2 \times 10^3$
Ketones	55 000	1850	2×10^4
	36 000	2750	$10 - 10^2$
Butadienes	46 000	2200	2×10^4
Nitroparaffins	37 000	2700	20

Table 12.2 *Electronic absorption frequencies of conjugated unsaturated acids in alcohol*

MOLECULE	POSITION (cm^{-1})	MOLAR EXTINCTION ε
$CH_3(CH=CH)COOH$	48 000	12 000
$CH_3(CH=CH)_2COOH$	37 000	25 000
$CH_3(CH=CH)_3COOH$	33 000	36 000
$CH_3(CH=CH)_4COOH$	30 000	49 000

hydroxyl and methyl groups, tend to lower the frequency and others to raise it. These two types are occasionally referred to as *bathochromic* and *hypsochromic* groups respectively. Table 12.3 shows the effect of methyl substituents on the lower absorption band of benzene.

Table 12.3 *Electronic absorption of methyl substituted benzenes*

COMPOUND	POSITION (cm^{-1})	MOLAR EXTINCTION ε
Benzene	39 200	230
Methyl benzene	38 300	300
1: 2 dimethyl benzene	37 600	415
1: 3 dimethyl benzene	37 600	400
1: 4 dimethyl benzene	37 600	780
1: 2: 3 trimethyl benzene	37 600	360
1: 2: 4 trimethyl benzene	36 400	835
1: 3: 5 trimethyl benzene	37 600	305
1: 2: 4: 5 tetramethyl benzene	35 800	820
Pentamethyl benzene	37 000	365
Hexamethyl benzene	36 800	300

A very large literature incorporating tables such as these is available, and much of it discusses the effect of small structural changes in some detail. Organic chemists regularly measure the spectra of unknown or uncertain compounds, especially in the fields of polyene, polyacetylene, aromatic, heterocyclic, steroid and carotenoid chemistry, to provide evidence for molecular structure. Even where the evidence is not of itself conclusive it may provide a guide as to what chemical reactions might prove profitable.

The effect of change of solvent has been widely studied, but this is not too important unless hydrogen bonding is present. Change of solvent may slightly vary the absorption frequency and also the precise shape of the bands and hence the extinction coefficient at a specified frequency. In quantitative analysis therefore calibrations and measurement should be carried out in the same solvent. In solution the bands may be from 200–8000 cm^{-1} broad and these are much greater than individual features in the gas or solid state. Presumably the fluctuations of the solvent in the neighbourhood of each molecule perturb the energy level of the excited state with its loosely bound electron. If there are discrete features in a liquid spectrum they relate to the possible different changes of vibrational energy which can be associated with the electronic transition. Since these frequencies in both the ground and excited states vary even between related molecules it is seldom that useful information is obtained from a qualitative examination of band shape. These changes of band shape are responsible for an uneven variation of ε_{max} between related molecules. The integrated extinction, $\int \varepsilon \, d\ln \nu$, is of greater theoretical significance but this quantity is less often evaluated.

Fully allowed transitions

While empirical correlations can be used for identifying organic structures it is naturally desirable to have a deeper understanding of the changes in electronic structure which are occurring. The excited states generally correspond to well-defined structures and energies and theoretical chemists have tried to match observations with calculations. This has proved to be a difficult task since the energy differences, which correspond to the observations, are small compared to the total energy of formation from separated atoms or from electrons and nuclei. The calculations may be simplified to semi-emirpical forms with greater success. The calculations also give an indication of the intensity of the transitions and the symmetry of the states involved. These can, with difficulty, often be measured or inferred from detailed consideration of rotational structure in the gas phase or polarization studies in crystalline solids.

As a result many of the simpler transitions have been identified. For many unsaturated compounds the low-lying levels correspond to the excitation of a bonding π electron to an antibonding π^* orbital. For aromatic compounds the ring plane is a nodal plane of all π orbitals which are compounded of atomic p orbitals whose axes are perpendicular to the ring. Such excitation leaves all principal quantum numbers unchanged. Because of this the excitation energy is only moderately high, $\sim$40 000 cm^{-1}, and the transition lies in the near ultra-violet or even the visible. They are often designated $\pi^* \leftarrow \pi$ transitions or more colloquially π to π *transitions*. For all these strong transitions the selection rule $\Delta S = 0$ holds and so, with spin paired ground states, they are singlet-singlet absorptions.

The detailed symmetries involved are described in terms of the group theory designations of the point groups and symmetry classes introduced in chapter 10. However, the idea behind the selection rules can be seen without these. Just as in the infra-red, the intensity is related to the square of the transition moment whose x component is indicated by $\int \Psi' p_x \Psi'' \, d\tau$, where Ψ' is the excited state wave function, Ψ'' the ground state function, p_x the dipole moment operator, i.e. Σex, and $d\tau$ is the volume element. In a reasonably good approximation for a one electron promotion this may be reduced to a one electron integral, $\int \varphi' x \varphi'' \, d\tau$, where the φ refer to one electron orbitals. While the detailed numerical evaluation requires specific forms, it can often be determined by inspection of the symmetry that certain of the integrals may be zero.

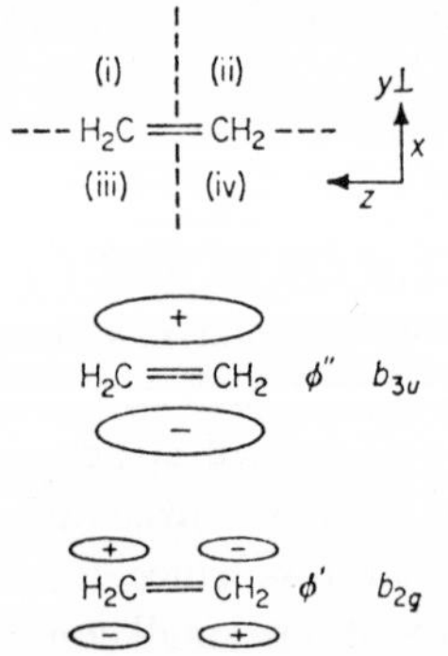

12.2 Edge-on view of ethylene indicating the subdivision into quadrants and the qualitative form of the filled bonding (b_{3u}) and empty antibonding (b_{2g}) one electron π orbitals.

Ethylene shows a simple case. Figure 12.2 shows at the top the co-ordinate system and below a representation of the phases of φ'' and φ'. The ethylene in the *yz* plane is viewed edge on. The molecule may be divided into four equivalent quarters (i) to (iv) and the integral over the whole of the volume can likewise be divided into integrals separately over these quarters. From the equivalence of the quarters each of these integrals will be equal in magnitude, though of either sign, for each of the operators x and z. Table 12.4 sets out the sign of each quantity and

Table 12.4

QUARTER	SIGNS OF φ'	φ''	x	z	$\varphi' x \varphi''$	$\varphi' z \varphi''$
(i)	+	+	+	+	+	+
(ii)	−	+	+	−	−	+
(iii)	−	−	−	+	−	+
(iv)	+	−	−	−	+	+
Sum					0	+4

See Figure 12.2 and main text.

their product for each quarter and indicates the sum over the four quarters. One can see that the integral $\int \varphi' x \varphi''\, d\tau$ is positive in two quarters and negative in the others leading to a total of zero. $\int \varphi' z \varphi''\, d\tau$ by contrast is positive in all four quarters and there is an allowed transition of intensity proportional to p_z where p_z the transition moment is $4e \times \int \varphi' z \varphi'' d\tau$, where the integral is over any individual quarter. The xz plane divides each quarter into two equal parts for one of which y is positive and for one negative. The φ are unchanged in passing through the xy plane and so $\int \varphi' y \varphi''\, d\tau$ is zero in each quarter. Consequently the transition is active only with z polarization.

Point group tables would indicate that the molecule was of symmetry D_{2h} (sometimes this is called V_h), that φ' belonged to the b_{2g} class, φ'' to b_{3u} and the product to b_{1u}. This class contains z, but not x or y, which is the condition that the transition be allowed and z polarized. For the specialist the use of group tables can be a saving of time and errors, but the method of Figure 12.2 and Table 12.4 leads to a readier understanding.

Vibrational structure

The energy of the transition is not always exclusively electronic, but some vibrational changes may also occur. For larger, flexible molecules in solution the structure is often blurred into a broad band, but in the gas or low temperature crystalline samples an elaborate vibrational structure may be apparent. The occurrence of particular vibrational intervals is related to the change of shape of the molecule on excitation, but usually the symmetrical skeletal vibrations are most prominent.

Electronic excitations may exist for which the electronic transition moments, $\int \Psi' x \Psi'' \, d\tau$, $\int \Psi' y \Psi'' \, d\tau$, $\int \Psi' z \Psi'' \, d\tau$, are all zero. Such transitions are forbidden in this approximation. They may in fact be observed but are commonly weaker by a factor of about 10^{-3}. They arise from a breakdown of the *Born–Oppenheimer approximation* in which the electronic and vibrational contributions are considered separately. In the analysis of the transition moment given for ethylene the molecule was assumed undistorted. If a distorted molecule had been considered, as it would need to be for a vibrating species, the contributions from each of the four quarters might not have been equal in magnitude and the detailed balancing of the $\int \varphi' x \varphi'' \, d\tau$ would not have occurred. Transitions which are dermitted solely on this account are called vibronic transitions or may be said to be *vibronically allowed.* Detailed quantum mechanical treatment can show which vibrational transitions must be invoked; a change in the vibrational quantum number of one such transition must occur on excitation. The pure electronic transition is strictly forbidden: this is often called the *0–0 band*, that is the band for which there is no vibrational quanta present in either the ground or the excited state.

The weak absorption of the $>C=O$ group near 36 000 is of this type, at least for symmetrical ketones. It arises from an excitation of a non-bonding lone pair electron in a p_y orbital into the antibonding π^*_x orbital. It can be called a $\pi^* \leftarrow n$ transition where n stands for non-bonding and especially a lone pair orbital.

Rydberg transitions and ionization

For saturated molecules all transitions occur at over 50 000 cm^{-1}, that is in the vacuum ultra-violet region where experimental difficulties are greater in view of the oxygen absorption. Many of these are broad and consist of overlapping features and have not received such intensive study.

There is one type of transition, analogous to an atomic transition, which has been recognized in the gas. It is the excitation of a valency electron into an orbital of large size. Since this electron is mostly at a considerable distance from the remainder of the molecule in the excited state, it tends to see the framework as if it were a structureless net positive charge of one unit. The form of the orbital of the excited electron resembles that in a hydrogen atom with a high principal quantum number and a series of transitions to such states are possible where the transition energies are of the form $A - Rhc/n^2$. Such series are called Rydberg series and converge to a limit at A, as $n \to \infty$, at which a continuum appears. Transitions at frequencies in the continuum correspond to ionization.

Near the continuum the lines of large n crowd together and the edge of the continuum may be difficult to detect. If the lower members of the Rydberg series can be identified, extrapolation to the ionization limit will be possible.

An alternative method of obtaining the ionization limit is to replace the detector behind the back slit with an ionization chamber containing the gas to be studied. As the frequency of the monochromator setting is increased, nothing is observed until the frequency of the continuum is just reached at which point ionization occurs and activates the chamber so that it conducts electricity.

Another important method recently introduced uses a monochromatic source, such as a xenon lamp, and estimates by how much the energy of the photons exceeds that required to produce ionization. The excess energy appears as kinetic energy of the ejected electron and this is readily measured by its ability just to pass through a grid with an approrpiate retarding potential.

Continua also occur when a molecule dissociates into two fragments. For excitation beyond the edge of the continuum the excess energy appears as kinetic energy of the fragments. Such fragments are usually free radicals with unpaired electrons and can be detected by the chain reactions they can initiate if, for example, they are allowed to impinge on a mixture of H_2 and Cl_2.

It is perfectly possible for excited molecules to have more energy than is required for ionization or dissociation and yet to remain intact. Such excited molecules return to a lower energy, usually on collision, in a time short compared to that required for the occurrence of that particular internal energy redistribution, which would lead to a large local excess energy in a particular bond and to its subsequent dissociation.

Lifetime of excited states

Studies of emission spectra in particular are influenced by the lifetime of the excited states and there are some features of this topic which are not yet fully understood. An excited molecule, assuming it does not react, can return to the ground state either by the emission of a quantum of radiation or by other mechanisms. It is then said to return by a *radiationless process* or by a *non-radiative transition*. Since for a given value of the transition moment, p_{ij}, the probability of spontaneous emission, A_{ij} (see p. 16), is proportional to ν_{ij}^3, such probabilities can get quite high for excited states which have fully allowed transitions to lower states in the visible or ultra-violet region. Though varying somewhat with circumstances a value of 10^{-8} s as the lifetime of such excited states is typical.

Competing, non-radiative, mechanisms may shorten the observed lifetime to values less than the natural lifetime. This lifetime limitation is also a limitation on the resolution of absorption spectra since the precision with which the energy of the excited state is known is limited by the Heisenberg uncertainty principle to $\hbar\tau^{-1}$ where τ is the lifetime. If

$\tau \sim 10^{-8}$ s the resolution, $\Delta\nu_{1/2} \sim 20$ MHz $\sim 10^{-3}$ cm^{-1}.

The radiative decay only involves a single molecule and the decay law is necessarily exponential, the fraction of excited states remaining after a time being exp $-t/\tau$. Most non-radiative decay processes are also first order in the excited state, and consequently exponential. The total rate constant is the sum of the rate constants. The lifetimes are the reciprocals of the rate constants and so

$$(\tau^{-1})_{\text{net}} = (\tau^{-1})_{\text{radiative}} + (\tau^{-1})_{\text{non-radiative}}.$$

The net lifetime is then somewhat less than the shorter of the separate lifetimes.

If the concentration of excited states is very high or the non-radiative contribution to the rate is small, the emitted light may be readily observed. The original excitation may be the result of a chemical reaction, an electric discharge or of an absorption process. Where emission follows absorption the term *fluorescence* is used. Since the emission is spontaneous it occurs in all directions. For most liquids and solutions at room temperature the non-radiative transitions are so rapid that fluorescence cannot be observed, but for some solutions, notably fluorescein, it is plain to the eye. Since the non-radiative processes are temperature dependent, cooling the solution, especially to supercooled liquids or glasses, often brings out the fluorescence spectrum. Aromatic impurities in crystals are also prone to strong fluorescence.

The detailed nature of the fluorescent spectrum does not match the absorption spectrum. The initial state of an absorption process is the ground vibrational state and the lowest frequency band is then the 0–0 band. The spectrum extends to higher frequencies when vibrational excitation as well as purely electronic excitation occurs, but lower frequencies are weak or absent; these lower frequencies correspond to 'hot' bands and appear when the temperature is high enough for a few vibrational states to be excited initially. The electronically excited molecule is usually able to establish thermal equilibrium between its vibrational states and so the fluorescence re-emission normally occurs from the lowest vibrational level. The 0–0 band is then the highest observed frequency and the emission extends to lower frequencies, the balance of the energy being retained by the molecule as vibrational energy. A study of the fluorescence spectrum may be a useful source of ground state vibrational frequencies. The energy diagram of Figure 12.3 may clarify

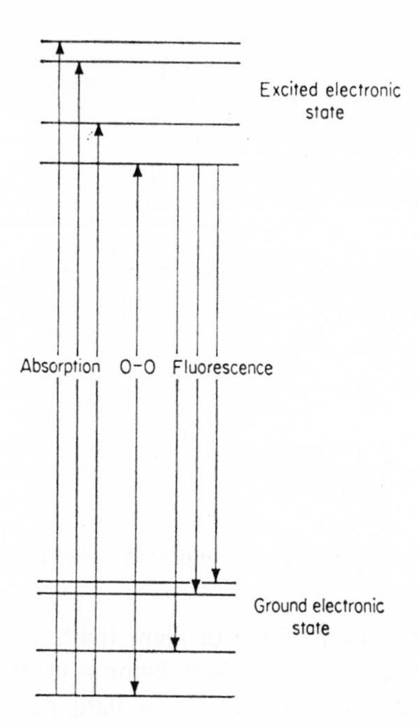

12.3 Energy diagram indicating absorption and fluorescence transitions. Three excited vibrational states are shown for each electron level.

the discussion. Figure 12.4 shows the absorption and emission spectrum of anthracene. If the vibrational frequencies and symmetries are similar in the two electronic states there is an approximate mirror symmetry to the absorption and fluorescence spectrum as in Figure 12.4.The 0–0 bands of fluorescence and absorption do not always occur at precisely the same frequency due to solvent effects. Absorption occurs with the solvent arranged in the suitable orientation for the ground state and fluorescence with the different solvation arrangement appropriate to the excited state. Since the absorption is at higher frequencies ultra-violet lamps are usually required to excite visible fluorescence.

If the molecule is in the gas phase at low pressure, the re-emission may occur from an excited vibrational level as the molecule is unable to share its vibrational energy with the surroundings by collision before re-emission. This case is called *resonance fluorescence*. Also if the molecule is in a solid and unable to rotate the emission may be partially polarized and emerge unequally at different angles with respect to the

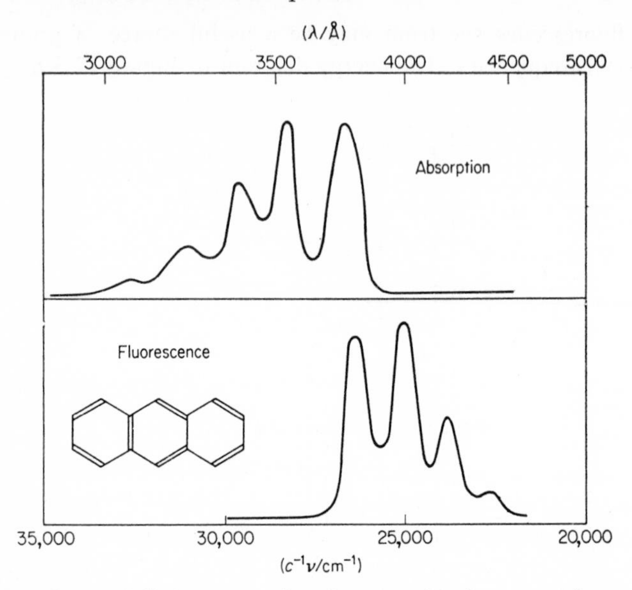

12.4 Absorption and fluorescence of anthracene showing approximate mirror symmetry.

direction and polarization of the incident light. There is thus a close connection with Raman spectra, there being a main line of unchanged frequency (compare Rayleigh line) with flanking bands on the low frequency side (compare Stokes lines), separated from it by the vibrational intervals of the ground state. Indeed, there is no sharp distinctive break in the series (a) Raman spectrum, (b) resonance Raman spectrum, (c) resonance fluorescence. Qualitatively these cases correspond to the incoming light being of a frequency at which the molecule is (a) transparent, (b) weakly absorbing in the wings of a band and (c) strongly absorbing. The general intensity of the emission varies (c)>(b)>(a) and there are differences in the relative strengths of the different features.

When associated with a fully allowed transition, the fluorescence is unlikely to have a radiative lifetime of more than 10^{-6} s, but for weakly allowed or vibrationally allowed transitions lifetimes of 10^{-4} s are possible. However, many cases of even longer decay times are known and times of seconds or even hours are possible. The name *phosphorescence* is usually applied in such cases. The difference between phosphorescence and fluorescence is largely a matter of degree, even though few decay times are known in the intermediate range, near 10^{-3} s. The long lifetimes, and consequential low values of A_{ij} and p_{ij}, mean that the emission is associated with a transition which is too weak to be easily seen in absorption. The nature of the transitions was a puzzle for many years until Lewis and Kasha suggested that they were spin forbidden triplet-singlet transitions. This view is certainly correct as shown by the detection of electron spin resonance in the triplet state of naphthalene which decays with the known phosphorescence lifetime. The electron spin selection rule, $\Delta S=0$, which forbids such transitions is weakened by small spin-orbit coupling terms in the quantum mechanical energy Hamiltonian. This same selection rule also influences the radiationless transitions so that the competing processes are also slow. Since the lifetimes are so long, the population of the triplet states can be made quite large and the phosphorescence is consequently not particularly weak. The $\Delta S=0$ selection rule is modified by collision with paramagnetic impurities and, in the presence of dissolved oxygen which itself has a triplet ground state, the absorption spectrum corresponding to the phosphorescence can just be observed at long path lengths. The lifetime of the phosphorescence is likewise reduced by the presence of oxygen, especially in mobile systems. At high concentrations of the triplet states one excited molecule is able to influence the lifetime of its neighbour and a bimolecular contribution to the decay is present. One such process is the formation from two triplet states in collision of one ground singlet and one excited singlet; the latter is capable of fluorescence. This situation is responsible for the delayed fluorescence whereby the singlet-singlet spectrum is emitted but after times of the order of magnitude normally associated with the phosphorescent spectrum.

A very different way to obtain large concentrations of excited states, and also of reactive dissociation products, is to use very large light intensities for the initial absorption. Such intensities are produced by discharging high capacity condensers, charged to a high voltage, through a gas discharge tube, often argon filled, which surrounds the main sample tube. Subsequent reactions can be followed by spectroscopic or other means. Especially when the primary process is photolytic decomposition the name *flash photolysis* is used to refer to this type of experiment. Most of the results refer to chemical reaction and chemical kinetics and are not of primarily spectroscopic interest. Sometimes the primary process may be excitation without decomposition and it may be possible to study further absorption by the molecules which are already excited electronically by the primary flash.

Optical activity, optical rotatory dispersion and circular dichroism

There is one special feature of the interaction of radiation and matter capable of electronic absorption which is of great importance in biochemistry and in natural product chemistry. This is the phenomenon of optical rotation in molecules of low symmetry. Indeed most optically active molecules have no symmetry, but some have a twofold axis; they may not have either a plane of symmetry or a centre of symmetry. Such optically active molecules, of which the tetrahedral CHFClBr would be a simple example, can in principle be separated into two types which are mirror images, I and II, and are referred to as D (for Dextrorotatory)

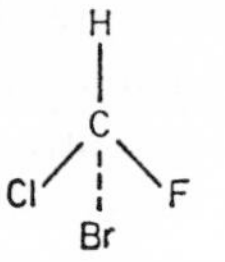

H, C, Cl, F, Br — I

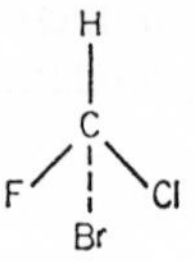

H, C, F, Cl, Br — II

and L (Laevorotatory) *enantiomorphs.* Such enantiomorphs are most easily distinguished experimentally by their ability, even in solution, to rotate the plane of polarization of polarized light which traverses them. One enantiomorph rotates the plane of polarization of light to the right, and the other to the left by the same amount at the same wavelength. Normally the D compound rotates the plane to the right for the yellow sodium doublet, but some compounds are named D and L from the structurally related key structures rather than by virtue of their own rotatory power. Laboratory syntheses produce inactive products from inactive materials, but nature herself has, over the centuries, developed a distinction and most natural materials occur in only one of the two possible optically active forms.

The ability to rotate the plane of polarization by a solution is directly related to different refractive indices for right and left circularly polarized light (see p. 13). Since these refractive indices are subject to anomalous dispersion, the refractive index differences increase as the light frequency approaches the transition frequency and even changes sign in the middle of a band in a manner reminiscent of Figure 2.2 on p. 21. A complete graph of optical rotation against frequency is called an *optical rotatory dispersion* curve and is more informative than a single measurement in the visible. In particular it may provide evidence for the molecular geometry in the vicinity of the absorbing group. Corresponding to the change with frequency of the real part of the refractive index difference, there is a difference of absorption coefficient for right and left circularly polarized light corresponding to the imaginary part of the refractive indices. As a consequence incident plane polarized light becomes eliptically polarized and the degree of ellipticity is a measure of the difference of absorption coefficients. A graph of this ellipticity against frequency is the *circular dichroism* curve. The complete circular dichroism curve and the complete optical rotatory dispersion curve are connected by a Kramers–Kronig (see p. 20) type relationship and they both contain the same essential information. At present the dispersion curve is more often measured, but the dichroism is gaining favour, especially where there is strong absorption. As might be expected the rotation and the dichroism are proportional to the concentration and to the path length in the solution, but may be expressed as a molar quantity like the extinction coefficients.

The detailed quantum mechanical expression of the phenomenon is somewhat involved. As has been seen, electrically allowed transitions are governed, as regards their intensity, by $(p_{ij})^2$ where (p_{ij}) is the electric dipole transition moment. The less common magnetically allowed transitions are similarly related to $(m_{ij})^2$ where (m_{ij}) is the magnetic transition moment. For the optically active transitions the scalar product $(m_{ij})(p_{ij})$ is the relevant quantity which governs the intensity of the optical activity effects. Although associated with absorption bands, it does not follow that the strongest absorption is associated with the strongest activity.

Brief mention must also be made of the *Faraday effect*. Molecules, not normally optically active, placed in a static magnetic field rotate the plane of polarization of radiation directed along the field: this direction would be along the axis of a solenoid or through the pole pieces of an electromagnet. The proportionality constants, called Faraday constants or Verdet constants, for visible light for most simple materials are known, but the variation of these constants with frequency near and in absorption bands has seldom been examined and should prove interesting.

13 Electronic spectra of diatomic molecules

Paradoxically the simplicity of diatomic molecules leads to a greater complexity in the electronic spectra observed compared to those of the larger polyatomic molecules. This is because the vibrational and rotational energy levels are fewer in number and transitions between identifiable, fully specified levels can be resolved with the best equipment and precise assignment is possible. For polyatomic molecules the rotational structure is frequently too finely spaced to be fully resolved and overlapping also occurs between vibrational bands; detailed assignment is not usually expected.

Experimental

To obtain a good absorption spectrum a long path length is required especially when the vapour pressure is low. Although modern folded path multiple transmission cells have alleviated this problem, it is as common to study the emission spectra to obtain the molecular information. There are two features which make emission spectra favourable for electronic transitions. One is that for the same transition moment and excited state population, the probability of spontaneous emission varies as the cube of the frequency (see p. 16). The second is that the usual detector, the photographic plate, is an inherent integrator and weak signals can be detected using exposure times of several hours or even days to obtain reasonable signal-to-noise ratios.

For moderate resolution a good glass or quartz prism spectrograph, such as that described for Raman investigations (Figure 9.1, p. 110), is suitable. For the highest resolution a close-ruled diffraction grating is superior. Some spectrographs have a plane grating and a collimating mirror, but others have the grating ruled on a spherical surface so that the one component acts both as a dispersing and as a focusing device as in

Figure 13.1. For work at 40 000 cm^{-1} the ruling would need to be of the same order, namely several thousand lines per cm. The resolving power is proportional to the total number of lines in the grating. The dispersion, that is the frequency spread per centimetre of plate, is also proportional to the distances involved, that is to the radius of curvature which may

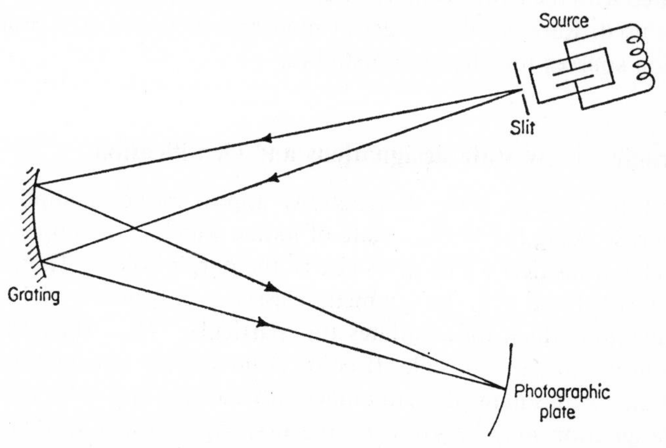

13.1 Simple form of high resolution visible or ultra-violet spectrograph.

be 3 metres or more. If the spectrum is not distributed over large lengths of the photographic plate in this way the grain size of the gelatine rather than the instrument becomes the factor which limits the resolution. The grating is set so that the required spectral range reaches the plate and the distances are adjusted for focus. The photographic plate may be based on thin glass which can be bent to fit a curved plate holder so that focus is maintained over its entire length. Besides convenience, there is an extra reason for simplicity in ultra-violet spectrometers, namely the poor reflecting power and high scattering by the metal surfaces which form the mirrors. A system with eight to ten reflections may transmit less than half the radiation.

The nature of the source depends strongly on the molecule or ion under investigation. A hot furnace may be used to raise the vapour pressure and a high voltage discharge can provide the electronic excitation. Photochemical reaction zones, shock tubes, and many other special sources are possible. Beyond 50 000 cm^{-1}, where absorption by oxygen sets in, evacuation of the whole spectrograph is common practice. Lithium fluoride is a good window material to 80 000 cm^{-1} (1200 Å) but beyond this no windows are effective and the emission should occur inside the vacuum system. For the absorption of non-corrosive gases the whole spectrometer can be filled with the sample gas. Absorption by the gelatine of the photographic plate is also a problem and the active silver bromide must be laid on the outer surface. Alternatively an efficient fluorescent dye may cover the surface and the ultra-violet light is thereby transformed to visible light, which travels through the gelatine and into the silver bromide grains. If a photomultiplier detector is required its window can be coated with the fluorescent dye and so normal glass or quartz envelopes can be used. Rare gas discharges at moderate pressures give continuous emission sources for absorption studies.

Electronic energy state designations and identification

At first the energy state designations appear rather complicated, a typical case being the $B^3\Pi_{0u+}$ state of iodine which is the upper state of the visible transition which gives rise to the purple colour. The initial B, not to be confused with the symmetry classes of polyatomic molecules, is a designation which differentiates this particular $^3\Pi_{0u+}$ state from any others in the iodine molecule. It conveys no specific information in the sense that a man's name does not inherently indicate any of his attributes. The designation may be given by the first experimenter to describe or classify the states and so far as possible the excited states are designated A, B, C, D, E, F in increasing order of energy; X is similarly used to denote the ground state. However, states may be unknown, be incorrectly described, relate to an impurity or ionic species or be simultaneously given different designations by different authors so that confusion is not unknown. The table in the back of *Spectra of Diatomic Molecules* by Herzberg is comprehensive to 1950 and is followed wherever appropriate. The left superscript, for example the 3 in $^3\Pi_{0u+}$, represents the multiplicity and has the same significance as in atoms, when Russell–Saunders coupling applies. It thus indicates $2S+1$, where S is the total electron spin quantum number. The possible values 1, 2, 3, 4 . . . , etc., are read singlet, doublet, triplet, quartet, etc. The central Greek letter indicates the value of the electronic orbital angular momentum about the interatomic axis in units of $\hbar$ and is therefore analogous to the Roman letters giving the orbital angular momentum for atoms. However, for linear molecules this momentum is constrained to be about the axis and, if not zero, it can have two directions, effectively clockwise and anti-clockwise, in contrast to the $2L+1$ directions for the angular momentum of atoms. Λ is the symbol for the associated quantum number analogous to L. The designations follow

Greek letter	Σ	Π	Δ	Φ	Γ	. . .
Value of Λ	0	1	2	3	4	. . .

which is the equivalent of the atomic table on p. 134. The electronic spin has a projection on the linear axis, usually designated Σ, not to be confused with the Σ of the preceding table, which may have the $2S+1$ integral values from $+S$ to $-S$. The sum $\Sigma+\Lambda=\Omega$ is written as the right subscript to the Greek letter. For the $^3\Pi$ states, Σ has the values $+1$, 0, -1, $\Lambda=1$ and $\Sigma+\Lambda=2$, 1 or 0 and the value 0 is seen to be appropriate for the $B^3\Pi_{0u+}$ state of I_2. The *g* (gerade or even) and *u* (ungerade or odd) symbols arise only with homonuclear molecules, that is when the two atoms are identical. They indicate whether the total electron wave function changes sign (*u*) or remains unchanged (*g*) on inversion of all the electron wave functions through the centre of symmetry between the nuclei. For a *Π* state the + sign refers to a feature of the coupling of the overall rotation and *Ω*. But for Σ states, when $\Sigma=0$, + or − is written as a right superscript which indicates if the electron wave function remains unchanged (+) or changes sign (−) on reflection through a plane containing the internuclear axis.

It can be seen therefore that a great deal of information is contained in a shorthand form in the state designations. This is particularly useful when applying the selection rules. The rule concerning *g* and *u*, namely only $g\leftrightarrow u$ transitions are allowed and the rule concerning + and − ($\Sigma^+\leftrightarrow\Sigma^-$ transitions only are allowed) are both strictly obeyed for changes involving radiation. The further rules $\Delta S=0$ and $\Delta\Lambda=0$, ±1 are less strict; they hold for light atoms but break down especially for heavy atoms with large spin-orbit couplings.

The determination of the full description and the associated quantitative features is a major part of this type of spectroscopy. The detailed rotational structure can be a valuable guide and so can the Stark and Zeeman effects in the presence of electrical and magnetic fields. Theoretical calculations can indicate the states to be expected in a particular energy range and help may be given by analogy with chemically similar molecules such as SnO and PbO.

Vibrational structure

Since the transitions are essentially electronic in origin, the vibrational contribution to the transition intensity multiplies the inherent electronic transition moment by the vibrational overlap integral, $\int \Psi'_{v'}\ \Psi''_{v''}\ d\tau$. But the different electronic states normally have different equilibrium internuclear distances and vibrational frequencies. Therefore the Hermite functions *Ψ*, which are the solutions to the vibrational problem, relate to different frequencies and different origins for the two electronic levels. This integral has no simple formulation and is non-zero for any value of

v' and v'', the vibrational quantum numbers. There is no strict vibrational selection rule therefore. For any specified values of v' and v'', when the states have vibrational frequencies ν' and ν'' and anharmonicity can be neglected, the change in vibrational energy is given by

$$\Delta E_{\text{vib}} = h\nu'(v' + 1/2) - h\nu''(v'' + 1/2).$$

This change must be added to the purely electronic change which is called the term value and written T_e. This is the difference between the equilibrium configurations which is not directly observable. A closely related quantity, ν_{00}, is the energy difference between the $v' = 0$, $v'' = 0$ states and such transitions are commonly, but not invariably, to be found in the observed spectrum. The difference lies in the zero point energy terms so that

$$T_e = \nu_{00} + (\nu'' - \nu')/2.$$

The formula for ΔE_{vib} shows that the spectrum will include sets of equally spaced bands at intervals of ν' which correspond to transitions with a common ν'' and likewise other sets separated by ν'' corresponding to transitions with a constant ν'. Such sets of bands are termed *progressions* in ν' and ν'' respectively. Only the lowest vibrational states are populated at equilibrium at room temperature and only ν' progressions are to be expected in absorption spectra. In emission thermal equilibrium is not maintained and the population of the separate v' levels may be dictated by conditions of the discharge or other factors.

Although the vibrational overlap integral quoted above is not easily evaluated, it is not an impossible task and tabulations are available. These confirm the reliability of the *Franck–Condon principle*. This important principle circumvents the exact quantum mechanical evaluation by taking a semi-classical viewpoint, which suggests that the absorption or emission of a photon is an instantaneous process and occurs in times too short for the heavy atomic nuclei to move significantly. Such transitions are termed vertical and correspond to the arrows of Figure 13.2. The two Morse curves there shown have different curvatures, different vibrational intervals and different equilibrium distances. For a classical oscillator in the $v' = 3$ level, the simple harmonic motion is such that the internuclear distance is usually that at one or other of the turning points of the motion. If emission occurs when it is at its inner turning point, the molecule is caught at a distance corresponding closely to the equilibrium of the ground state and the $3 \rightarrow 0$ vibrational change is expected to occur. If the oscillator is at the outer turning point of the motion with $v' = 3$, change of electronic energy without change of internuclear distance leaves the molecule near the outer turning point of the $v'' = 7$ and $v'' = 8$

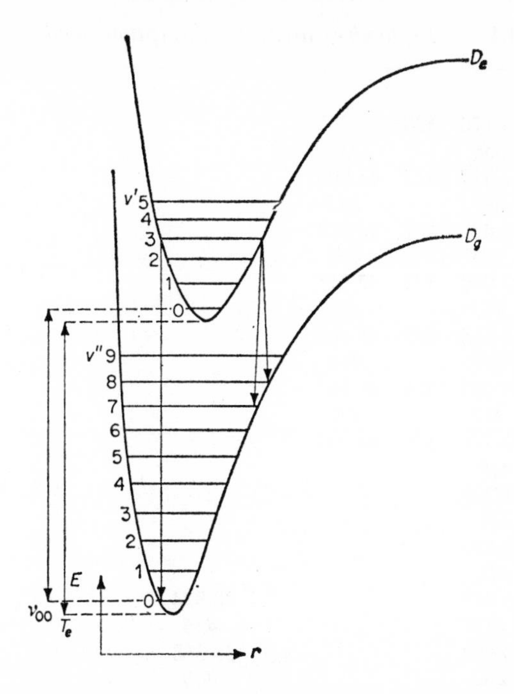

13.2 Typical ground and excited state Morse curves. Most probable emission transitions from the $v' = 3$ state are shown.

states. The Franck–Condon principle then states that for this particular example emission from the $v' = 3$ level will occur predominantly to the $v'' = 0$, 7 and 8 states. Quite weak transitions to the $v'' = 1$, 6 and 9 states are to be expected since the principle merely identifies the strongest transitions, but transitions from $v' = 3$ to the $v'' = 2$, 3, 4, 5, 10, 11 . . . states are not favoured. The principle also applies to polyatomic molecules, where it may be involved in each vibrational degree of freedom. However, for large polyatomic molecules large changes of frequency or internuclear distance are less common and there are seldom more than one or two vibrations for which Δv is not equal to zero for the more intense transitions.

When anharmonicity is considered the vibrational intervals are not equal, yet each interval may appear several times associated with different vibrational energy in the other state. If the transition frequencies are set out in a rectangular array, as in Table 13.1, these vibrational intervals appear as the difference between adjacent numbers and the assignment can be readily inspected. Such a presentation is called a *Deslandres table*.

Table 13.1 *Deslandres array of absorption bands of SnO*

v' \ v''	0		1		2		3		4
18	44 170	*818*	43 352						
	358		*352*						
17	43 812	*812*	43 000						
	387		*382*						
16	43 425	*807*	42 618						
	395		*403*						
15	43 030	*815*	42 215						
	412		*409*						
14	42 618	*812*	41 806						
	415		*419*						
13	42 203	*816*	41 387						
	427		*430*						
12	41 776	*819*	40 957						
	440								
11	41 336								
	441								
10	40 895								
	448								
9	40 447				38 824				
	455				*456*				
8	39 992				38 368				
	464				*460*				
7	39 528	*814*	38 714	*806*	37 908	*800*	37 108		
	471		*469*						
6	39 057	*812*	38 245						
	472		*472*						
5	38 585	*812*	37 773						35 374
	477		*480*						
4	38 108	*815*	37 293	*807*	36 486	*801*	35 685		
	479		*485*		*483*				
3	37 629	*821*	36 808	*805*	36 003				
	490		*485*		*490*				
2	37 139	*816*	36 323	*810*	35 513	*803*	34 710	*789*	33 921
	495		*498*		*496*		*494*		*495*
1	36 644	*819*	35 825	*808*	35 017	*801*	34 216	*790*	33 426
					501		*502*		*504*
0					34 516	*802*	33 714	*792*	32 922
v' / v''	0		1		2		3		4 v''

Wave numbers in cm^{-1}. Vibrational intervals are given in italics.

Rotational structure

The wave functions associated with the molecular rotation are surface harmonics which depend only on angles and not on the internuclear distance. These appear as factors of the total wave function and the associated transition intervals are the same for all linear molecules without electronic angular momentum; the selection rule is $\Delta J = \pm 1$. $\Delta J = 0$ transitions are also allowed when there is electronic orbital angular momentum around the axis.

For a transition with $\Delta J = +1$, $J' = J'' + 1$ and

$$\Delta E_{\text{rot}} = hB'(J''+1)(J''+2) - hB''J''(J''+1) = h(J''+1)(B'+B'') + h(J''+1)^2(B'-B'');$$

J'' may be zero or any positive integer. For $\Delta J = -1$, $J' = J'' - 1$ and

$$\Delta E_{\text{rot}} = hB'(J+1) - hB''(J+1)(J'+2) = -h(J'+1)(B'+B'') + h(J'+1)^2(B'-B''),$$

where J' may be zero or any positive integer. These two expressions can be combined to give

$$\Delta E_{\text{rot}} = hJ(B'+B'') + hJ^2(B'-B''),$$

where J may be any positive or negative integer but not zero. If only the first term were present the rotational structure would consist of a set of evenly spaced lines with frequency separation $(B'+B'')$ except that the central line would be absent as $J \neq 0$. However $(B'-B'')$ may not be zero. If this bracket is negative the lines with $J<0$ have a progressively wider spacing and those with $J>0$ crowd together and eventually a maximum value of ΔE_{rot} is obtained. Such bands appear to have a sharp cut-off or band head on the high-frequency side and are said to be degraded to low frequencies or degraded to the red. Such spectra occur if $B'<B''$ or $I'>I''$ and the equilibrium separation is greater in the excited than in the lower state. This is the most common case, but if the internuclear distance shortens on excitation, the band head is on the low-frequency side and the band degrades to the blue. The direction of the degradation can usually be obtained with spectrometers of modest resolution even if the individual rotational lines are not resolved.

Further complexities arise when $\Delta J = 0$ transitions are present and form a strong central Q-branch and when centrifugal distortion and similar features are included.

Dissociation

It may happen that a molecule absorbs from the radiation more energy than that required to dissociate the molecules into two atoms. However, such dissociation does not necessarily ensue. For instance if Figure 13.2 represents the potential energy curves, absorption could take place to the levels $v' = 4$, 5, etc., without dissociation even though the total energy is greater than D_g, the ground state dissociation energy. The excited molecule could re-emit the radiation and return directly to the ground state. It could also lose vibrational energy falling to the $v'=0$ level, which also re-emits radiation when the molecule returns to the ground electronic states. There is also a slight possibility of a radiationless transition from the $v'=0$ state to a high vibrational state, about $v''=11$ of the ground electronic state; this process is more probable in condensed media than in a gas at low pressure.

Absorption direct to a higher energy in the excited state, that is above D_e, does lead to dissociation directly. One product of the dissociation process in this case is an atom in an excited electronic state, with excitation energy $(D_e - D_g)$. Less often this excess energy is shared between the two atomic products. Thus the visible absorption band of I_2 has a set of vibrational bands leading into a continuum corresponding to the dissociation of the $^3\Pi_{0u+}$ excited state into a ground state atom, $^2P_{3/2}$ and an excited atom in the $^2P_{1/2}$ level. The ground molecular state dissociates into two ground state atoms, $^2P_{3/2}$ so that $(D_e - D_g)$ is equal to the excitation energy $^2P_{1/2} \leftarrow {}^2P_{3/2}$, namely 7603 cm^{-1}.

The absorption spectrum in such a case consists of a series of sharp bands, corresponding to low values of v', which break off (at about $v'=11$ for the diagram of Figure 13.2) to be replaced by a continuum at high frequencies. This continuum corresponds to the arbitrary amount of energy taken away as kinetic energy of the two atomic fragments. The onset of this continuum gives D_e referred to the $v''=0$ level, and if the atomic excitation can be identified so that $(D_e - D_g)$ is known, the ground state dissociation energy, D_g, can be deduced.

Sometimes absorption near the continuum is weak, but the position of the edge may be found by extrapolation of the vibrational intervals. Because of anharmonicity the vibrational intervals become less at high v' and finally vanish at the dissociation. If the lower levels are represented well by

$$E = h\nu'(v'+1/2) - h\nu'x'(v'+1/2)^2,$$

the interval vanishes at $(v'+1/2) = 1/2x'$ and the energy at this point is $h\nu'/4x'$. This case with only the first anharmonic coefficient, x', corresponds exactly to a Morse curve but more complex energy formulae can be handled similarly. Graphical methods are also used. This energy refers to the minimum of the excited electronic state; the term value T_e must be added and the atomic excitation, $(D_e - D_g)$, subtracted to give the ground dissociation energy D_g. In principle extrapolation of the ground state vibrational intervals would also give D_g, but levels higher than $v''=0$ or 1 are not common in absorption at room temperature and high values of v'' are not always prominent in emission. Also the ground state curve is usually deeper, the anharmonicity consequently smaller, and the answers then become less accurate as the extrapolation must be made over a greater energy range.

If the potential energy curves of importance are arranged as in Figure 13.3 all the absorption processes lead to dissociation and only a con-

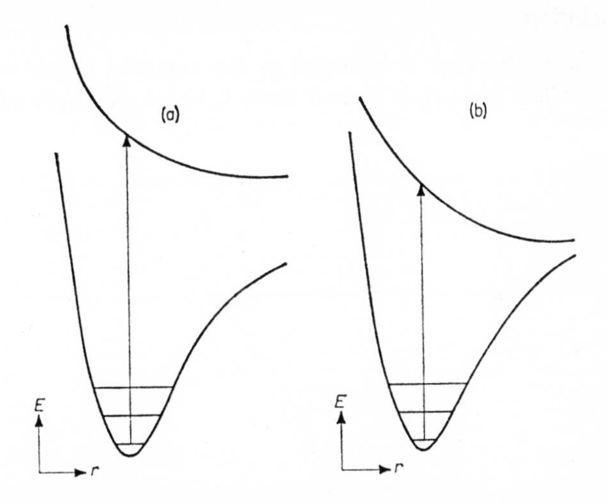

13.3 Potential energy curves with dissociating upper states. (a) Products include at least one excited atom, (b) products are ground state atoms.

tinuous spectrum without vibrational features can be observed. The upper curves of Figure 13.3 are repulsion curves with no stable minimum. They correspond really to the potential energy of a non-associative collision between two atoms, which experience a steep repulsion when they are at a distance less than the sum of their van der Waals' radii. The onset of the continuum, that is the lowest possible absorption energy, corresponds to the energy to form two atoms which may (Figure 13.3(b)) or may not (Figure 13.3(a)) be in their ground atomic states. The most intense part of the continuum corresponds, in accordance with the Franck–Condon principle, to vertical transitions, such as those indicated. These occur at considerably higher frequencies than the edge of the continuum. This edge can be extremely difficult to observe accurately on a photographic plate. Indeed, it may be more profitable to use some method of detecting atoms, such as their ability to initiate chain reactions; in this way the minimum frequency required for the production of atoms may be detected with greater sensitivity. The case of Figure 13.3(a) is commoner than that of Figure 13.3(b). Where two curves correspond to the same atomic fragments they usually correspond to different spin multiplicities and electronic transitions between them are then weak or forbidden. Thus levels of hydrogen are disposed as in Figure 13.3(b), but the repulsion curve corresponds to a triplet state whereas the attractive ground state is a singlet.

Predissociation

Another interesting case is indicated in the potential energy curves of Figure 13.4. The immediate excited state, I, has a stable minimum, but

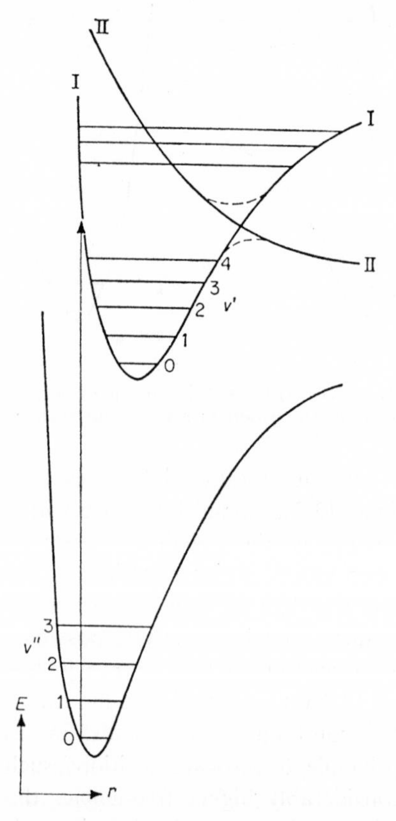

13.4 Potential energy diagram showing predissociation.

it is crossed by a dissociative state, II. In a first approximation these states may be considered separately and if there are allowed transitions between the ground state and I, a banded spectrum might be expected. However, exact calculations including any small element in the Hamiltonian which mixes states I and II lead to the operation of the non-crossing rule (see p. 4) and the potential energy curves, appropriate to static nuclei, follow the dashed curves near the crossing point. The vibrational levels below the crossing point ($v' \leqslant 4$ in Figure 13.4) remain well quantized and transitions to these levels give the usual sharp bands. So too do transitions to the levels well above the crossing point. This is less expected at first sight, as there is more than sufficient energy to cause dissociation. However, if the absorption of radiation leaves the system with energy nearly equal to that at the crossing point, the system is in a state compounded from wave functions appropriate to both curves I and II and the likelihood that the two atoms will follow II and dissociate is quite high. The lack of quantization of the relative kinetic energy of the two atoms means that there is no precise energy requirement and the absorption spectrum is continuous. The exact calculations of the probability of dissociation is quite troublesome, but there is a requirement, expressed in classical language, that the probability of dissociation is high if the system has very little kinetic energy when it is near the crossing point. One may make an analogy with a straight putt on a golf green: if the ball's kinetic energy is low when it reaches the hole it will roll in, but if the velocity is too high the ball will cross over the top of the hole and continue its path on the far side.

The net absorption for such a case consists of sharp vibrational bands at low frequencies, a continuum region at intermediate frequencies and sharp bands again at higher frequencies merging into a second, normal type of dissociation continuum when the energy is above the asymptote to curve I. In this sense the dissociation occurs before or 'pre' the regular position, when such dissociation is associated with the crossing point. The vertical or Franck–Condon absorption indicated by the arrow in Figure 13.4 corresponds to dissociation; classically the molecule would make half a vibration before dissociation. It can be seen that vertical transitions direct to state II would be at much higher frequencies. If the probability of crossing to state II is very low the spectrum may not be truly continuous, but consist of vibrational bands with poorly resolved or continuous rotational fine structure. Various subdivisions of the types of predissociation are made according to the details, such as whether curve II also has a minimum, etc., but the cases are all based on the crossing of energy levels.

Problems

The following problems are based on the chapters indicated by the numerals which follow them. Some are fairly difficult. They have been devised to amplify points in the text and to give the serious student a chance to judge his own understanding. Some solutions are given on p. 185.

1. What is the population at thermal equilibrium at 300 K relative to the ground state as unity of
(i) a non-degenerate nuclear quadupole state at 100 MHz
(ii) a gas rotational energy level of sevenfold degeneracy at 30 cm^{-1}
(iii) vibrational states at 384 cm^{-1} and at 1800 cm^{-1}
(iv) an electronic state at 40 000 cm^{-1}? 1

2. Write out in full the derivation of

$$dn_i/dt = \tau^{-1}(n_i^{\infty} - n_i)$$

sketched in the text on p. 6. 1

3. To the wave functions of p. 18 add a fourth $\Psi_{J=3} = 5\cos^3\theta - 3\cos\theta$ and determine all (unnormalized) transition moments p_{ij} for all transitions with $J', J'' \leqslant 3$ including the cases $J' = J''$ and verify the selection rule $\Delta J = \pm 1$. 2

4. A transition has $p_{ij} = 3{\cdot}3 \times 10^{-30}$ m s A at 30 cm^{-1}. Calculate:
(i) the natural line width
(ii) the radiation induced line width at 300 K
(iii) the collision line width for $\tau = 10^{-10}$ sec
(iv) the Doppler line width at 300 K if the molecular mass is 5×10^{-26} kg
(v) the total line width. 2

5. Plane polarized radiation of 5000 Å wavelength traverses 1 cm of a solution of optically active material and suffers a clockwise rotation of $+7{\cdot}2°$. If the mean refractive index is 1·40 calculate the refractive index difference of the solution between right and left circularly polarized radiation. 2

6. Draw a schematic energy level diagram for CH_2F_2 spin states and label these with values of I and M_I for each type of nucleus. Indicate the allowed transitions, obtain their frequencies and indicate the spectra to be expected. Assume $\nu_H = 42$ MHz, $\nu_F = 40$ MHz, $J_{HH} = -12$ Hz, $J_{HF} = +60$ Hz and $J_{FF} = +160$ Hz. 3

7. $CF_2{=}CFCl$ gives a fluorine resonance spectrum at 30 MHz of twelve lines of about equal intensity with frequencies (decreasing in field) of 0, 58, 115, 173, 692, 770, 807, 885, 1205, 1263, 1283 and 1341 Hz referred to an arbitrary zero. Suggest an interpretation of this spectrum. 3

8. The maximim splitting, for any orientation, in the gypsum spectrum (Figure 3.11, p. 41) is 2·16 mT. Assuming an O–H bond length of 0·98 Å calculate the HOH angle in the water of crystallization. 3

9. Figure P.1 shows the high resolution nuclear resonance spectrum at

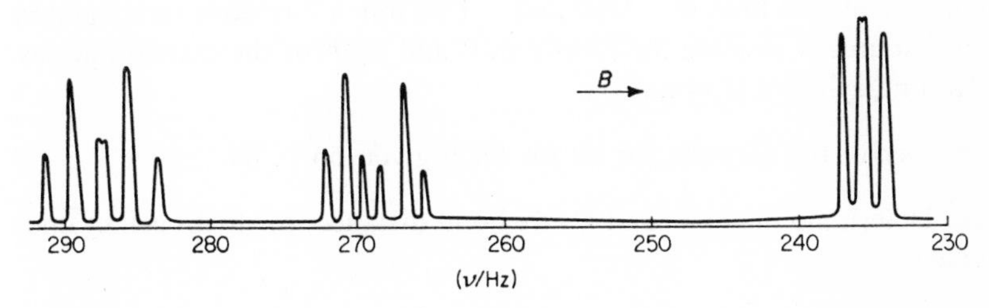

P.1 60 MHz high resolution nuclear magnetic resonance spectrum of diketene.

60 MHz of diketene. The exact line positions are at fields equivalent to the following frequencies measured downfield from tetramethyl silane as origin:

291·18, 289·36, 287·54, 287·26, 285·44, 283·62, 272·15, 270·81, 269·47, 268·28, 266·89, 265·55, 237·18, 235·84, 235·36 and 234·02 Hz.

Interpret the pattern in terms of τ values and coupling constants for the hydrogen atoms. 3

10. $h^{-1}eqQ$ for a p_z orbital of atomic iodine is $-22\,292{\cdot}89$ MHz. Using the figures in the main text, p. 47, derive U_p for the bonds of SnI_4. 4

11. At 20 K two quadrupole resonances are observed for tertiary butyl chloride, $(CH_3)_3CCl$, the stronger at 31·195 MHz and the weaker at 24·586 MHz. Interpret these observations. 4

12. Liquid ethane, C_2H_6, in the cavity of an electron resonance spectrometer working at 9240 MHz was bombarded with 2·5 MeV electrons. The magnetic field was swept slowly and twelve lines of approximate relative intensities 1:2:3:1:6:3:3:6:1:3:2:1 were observed at 321·98; 324·23; 324·69; 326·48; 326·94; 327·40; 329·19; 329·66; 330·11; 331·91; 332·35; 334·61 mT respectively. Interpret these observations. 5

13. For a given orientation of the radical $CH(OH)CO_2^-$ the two hydrogens have effective hyperfine couplings of -64 MHz and $+40$ MHz. Draw a schematic energy level diagram of the style of Figure 5.7 on p. 59 for this radical in a magnetic induction of about 325 mT. Take the electron resonance frequency to be 9100 MHz and the normal hydrogen nuclear resonance frequency to be 14 MHz in this field. Indicate the allowed electron resonance transitions $\Delta M_S = 1, \Delta M_I = 0$ and the allowed nuclear resonance frequencies, $\Delta M_S = 0, \Delta M_I = 1$ for one nucleus and $\Delta M_I = 0$ for the other. Tabulate the frequencies. 5

14. The Mössbauer spectrum of ^{57}Fe in ferrocene, $Fe(C_5H_5)_2$, at 20 K consists of two lines at $-0·50$ and $+1·88$ mm s^{-1} relative to a stainless steel source. Calculate the isomer shift and eqQ for the excited nucleus. Assume cylindrical symmetry. 6

15. Derive the formula for ΔI for OCS given on p. 88. 7

16. Determine the H–C and C≡N bond distances for hydrogen cyanide given

$^1H^{12}C^{14}N$ $B_0 = 44\,316$ MHz.
$^2D^{12}C^{14}N$ $B_0 = 36\,208$ MHz. 7

17. Deduce the formula for rotational transition frequencies of a linear molecule when D_J is not neglected. Fit the data of Table 7.1 on p. 77 to this formula and deduce B_0 and D_J for $^{16}O^{12}C^{32}S$.

18. $c^{-1}B_0$ for $^{69}Ga^{35}Cl$ is 0·1486 cm^{-1} and the vibration-rotation interaction constant α is $+0·0004$ cm^{-1}. Determine r_e.
(Atomic masses on $^{12}C = 12$ scale $^{69}Ga = 68·93$, $^{35}Cl = 34·97$.) 7

19. The table below gives ε'' for toluene ($C_6H_5CH_3$) at 293 K. Show that toluene obeys the Debye dielectric loss curve for a single relaxation time within experimental error. Determine $(\varepsilon_S - \varepsilon_\infty)\varepsilon_0^{-1}$, τ and p, the dipole moment. $\varepsilon_S\varepsilon_0^{-1} = 2·3$, density $= 0·86$ Mg m^{-3} both at 293 K.

$\varepsilon_0^{-1}\varepsilon''$	0·038	0·060	0·058	0·058	0·048 ± 0·002
7ν	9·2	18	24	33	50 × 10^9 Hz

20. A mixture of 2-, 3- and 4-methyl heptanes (C_8H_{18}) in a cell of thickness 0·13 mm had optical densities of 0·45 at 1170 cm^{-1}, 0·18 at 770 cm^{-1} and 0·97 at 748 cm^{-1}. The molecular extinction coefficients, ε, are given in the table. The density of each component and the mixture may be taken as 0·82. What are the proportions of each isomer? 8

Wave number	1170	770	748 cm^{-1}
Extinction coefficient ε for			
2-methyl heptane	11·8	2·1	3·2
3-methyl heptane	2·1	5·2	3·6
4-methyl heptane	3·8	0·9	14·4

21. The Raman spectrum from liquid chloroform excited by the 4358·3 Å line of mercury consists of lines at 4289·9, 4309·1, 4358·3, 4408·7, 4429·0, 4489·1, 4507·9, 4602·2, 5018·7 Å. What are the six fundamental frequencies disclosed in units of cm^{-1}? 9 and 10

22. The $^{16}O_2$ molecule has $c^{-1}\nu_e = 1580\ cm^{-1}$, $xc^{-1}\nu_e = 12{\cdot}1\ cm^{-1}$ and $r_e = 1{\cdot}207$ Å. Calculate the bond stretching force constant. Calculate the wave numbers of the v 1↔0, 2↔1, 3↔2 and 2↔0 transitions. Assuming the molecule exactly follows the Morse potential function obtain a value for the dissociation energy D_e. How many vibrational levels lie below the limit? The accurate D_e is 3·484 kJ mol^{-1} (41 770 cm^{-1}). Using this figure and assuming a Morse potential function obtain an estimate of $xc^{-1}\nu_e$. 10

23. Sketch a figure like Figure 10.5, p. 123, but showing the opposite phase of the vibrational motions. 10

24. What is the net energy in kJ mol^{-1} evolved in the reaction

$$HD + HCl \rightarrow H_2 + DCl$$

if all the molecules are in their ground vibrational states? $c^{-1}\nu(H_2) = 4395$, $c^{-1}\nu(HD) = 3817$, $c^{-1}\nu(HCl) = 2990$, $c^{-1}\nu(DCl) = 2091\ cm^{-1}$. 10

25. A high-resolution Raman spectrum of N_2 excited by the 4358 Å line of mercury gave lines at the following wave numbers: 22 988·54, 22 980·54, 22 972·53, 22 964·53, 22 956·53, 22 944·52 (very strong), 22 932·51, 22 924·51, 22 916·51, 22 908·50, 22 900·50, 22 892·48, 22 884·48, 20 658·22, 20 650·07, 20 641·95, 20 633·87, 20 625·83, 20 614 (strong and broad), 20 601·93, 20 594·04, 20 586·19, 20 578·37, 20 570·60, 20 562·86, 20 555·16, 18 312 (broad) cm^{-1}. Determine B_e, α, r_e, $c^{-1}\nu_e$, $xc^{-1}\nu_e$ and k from these results. Mass of $^{14}N = 14{\cdot}0038$ ($^{12}C = 12$). 9 and 10

26. Consider the shell filling in the ground state of phosphorus (15 electrons) and sulphur (16) atoms. Suggest the probable designation of the ground state with Russell–Saunders coupling. 11

27. In addition to those in Table 11.2, p. 143, some weaker lines of the sodium spectrum can be observed at the following wave numbers in cm^{-1}.

5867·4	9300·4	12 800·5	39 794·8
5869·9	9302·9	12 801·2	40 137·2
8123·6	11 556·6	19 398·4	
8124·9	11 557·9	19 415·5	

Suggest assignments for these lines and consequent new entries for Table 11.3. 11

28. Figure P.2 shows the appropriate phases of the atomic orbitals of the one electron orbits of cis-butadiene. The ground state shell filling is

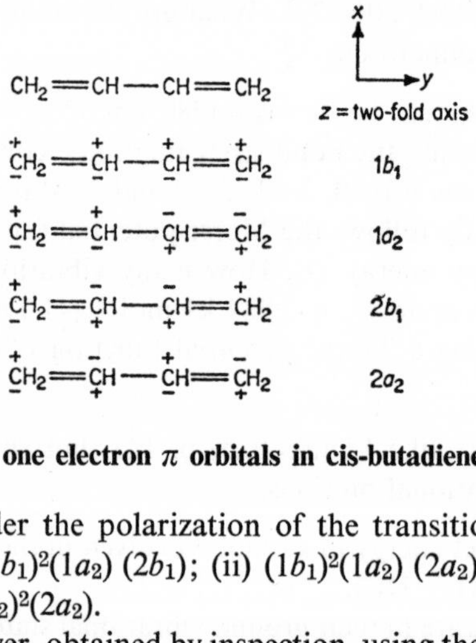

$CH_2{=}CH{-}CH{=}CH_2$

x, y

z = two-fold axis

$CH_2{=}CH{-}CH{=}CH_2$ (above: + + + +; below: − − − −) $1b_1$

$CH_2{=}CH{-}CH{=}CH_2$ (above: + + − −; below: − − + +) $1a_2$

$CH_2{=}CH{-}CH{=}CH_2$ (above: + − − +; below: − + + −) $2b_1$

$CH_2{=}CH{-}CH{=}CH_2$ (above: + − + −; below: − + − +) $2a_2$

P.2 Symmetries of one electron π orbitals in cis-butadiene viewed edge on.

$(1b_1)^2(1a_2)^2$. Consider the polarization of the transitions to the singlet excited states (i) $(1b_1)^2(1a_2)(2b_1)$; (ii) $(1b_1)^2(1a_2)(2a_2)$; (iii) $(1b_1)(1a_2)^2(2b_1)$; (iv) $(1b_1)(1a_2)^2(2a_2)$.

Confirm the answer, obtained by inspection, using the C_{2v} group multiplication tables $a \times a = b \times b = a$, $a \times a = b$, and $1 \times 1 = 2 \times 2 = 1$, $1 \times 2 = 2$ for subscripts, given that x is in b_1, y in b_2 and z in a_1. 12

29. The following assignments have been suggested for vibrational bands in the absorption spectrum of AgI near 3200 Å. The wave number in cm^{-1} is followed by the assigned values of $(v' \leftarrow v'')$ in parenthesis.

31 634 (5←0), 31 569 (4←0), 31 482 (3←0),
31 384 (2←0), 31 272 (1←0), 31 154 (0←0),
30 948 (0←1), 30 744 (0←2), and 30 540 (0←3).

Suggest assignments for bands at 31 066, 30 861, 31 177, 30 769, 30 868 cm^{-1}. At what wave numbers should the (2←2), (1←3), (3←1) bands be sought? Show that the anharmonicity in the excited state does not fit the expression derived from a Morse curve.

Answers to problems

1. (i) 0·999 984 (ii) 6·1 (iii) 0·160 and $1{\cdot}8 \times 10^{-4}$ (iv) 8×10^{-84}.

3. $0 \leftrightarrow 1$, $2p/3$; $1 \leftrightarrow 2$, $8p/15$; $2 \leftrightarrow 3$, $24p/35$; others 0.

4. (i) 5·6 MHz (ii) 490 MHz (iii) 1600 MHz (iv) 1·0 MHz (v) 1680 MHz ≡ 0·056 cm^{-1}.

5. $1{\cdot}0 \times 10^{-6}$.

6. Energy levels:

	M_F	M_H	MHz	+Hz		M_F	M_H	MHz	+Hz
a	−1	−1	82	97	i	0′	0′	0	−111
b	0	−1	42	37	j	0′	0	0	−123
c	0′	−1	42	−123	k	−1	1	−2	−23
d	−1	0′	40	49	l	1	0′	−40	49
e	−1	0	40	37	m	1	0	−40	37
f	1	−1	2	−23	n	0	1	−42	37
g	0	0′	0	49	o	0′	1	−42	−123
h	0	0	0	37	p	1	1	−82	97

(0′ = antisymmetric component)

Allowed transitions:

MHz	+Hz	
42	60	a↔e, e↔k
42	0	b↔h, c↔j, h↔n, j↔o
42	−60	f↔m, m↔p
40	60	a↔b, b↔f
40	0	d↔g, e↔h, g↔l, h↔m
40	−60	k↔n, n↔p

7. δ_1 −424·3, δ_2 −266·2, δ_3 −2·9 ppm w.r.t origin. $|J_{12}|$ 78, $|J_{13}|$ 58, $|J_{23}|$ 115 Hz.

8. 108°.

9. τ_1 3·19, τ_2 2·99, τ_3 2·62; $|J_{12}|$3·92, $|J_{13}|$1·82, $|J_{23}|$1·34 Hz. 1,2 cis/trans exocyclic, 3 two fully equivalent ring hydrogens.

10. 0·585 axial; 0·592 others.

11. ^{35}Cl and ^{37}Cl resonances, $|h^{-1}eqQ|$62·39 and 49·17 MHz respectively, $Up = 0{\cdot}573$.

12. CH_3CH_2 radical, $g = 2{\cdot}0037$, $|a_{CH_3}|$ 75·9, $|a_{CH_2}|$ 63·0 MHz (2·71 and 2·25 mT).

13. Level energies, 4576, 4570, 4530, 4524, −4524, −4542, −4558, −4576 MHz. Transitions, 9152, 9112, 9088, 9048; 46, 18; 34, 6 MHz.

14. Shift 7·9 MHz, $|h^{-1}eqQ|$55 MHz.

16. 1·07, 1·15 Å.

17. $\nu_{J \leftarrow J-1} = 2BJ - 4D_J J^3$; 6081·49 MHz, 1·309 kHz.

18. 2·213 Å.

19. 0·119; 8×10^{-12} s; $1{\cdot}1 \times 10^{-30}$ m s A.

20. 0·18, 0·18, 0·64 respectively.

21. 262, 366, 668, 761, 1216 and 3019 cm^{-1}.

22. 1170 kg s^{-2}, 1556, 1531, 1508, 3087 cm^{-1}; 4·30 kJ mol^{-1} (51 600 cm^{-1}); 65; 14·9 cm^{-1}.

24. 1·92 kJ mol^{-1}.

25. 2·010 cm^{-1}, 0·0187 cm^{-1}, 1·094 Å, 2359·6 cm^{-1}, 14·46 cm^{-1}, 2300 kg s^{-2}.

26. $(1s)^2(2s)^2(2p)^6(3s)^2(3p)^3$, $^4S_{3/2}$; $(1s)^2(2s)^2(2p)^6(3s)^2(3p)^4$, $^3P_2$.

27. 5p↔3d, 5p↔3d, 6p↔3d, 6p↔3d, 5p↔4s, 5p↔4s, 6p↔4s, 6p↔4s, 7p↔4s, 7p↔4s, 6s↔3p, 6s↔3p, 9p↔3s, 10p↔3s
6s 36 371·1 cm^{-1}, $^2S_{1/2}$, 0·2151, 4·649, 1·291
9p 39 794·8 cm^{-1}, ^2P_u, 0·1228, 8·146, 1·105
10p 40 137·2 cm^{-1}, ^2P_u, 0·1094, 9·144, 1·094.

28. y, z, z, y.

29. 1←1, 2←3, 3←3; 31 177, 30 667, 31 275 cm^{-1}.

Bibliography

The following guide to further reading covers principally authoritative books available in English, although a few review articles are included. In some areas only one book of authority may have been written while elsewhere the field is large and there are other books of comparable merit which might have been included, except that some attempt has been made to cover the ground evenly. In the immediately following paragraphs the works are referred to by their authors' names; and detailed titles are given subsequently.

For an introduction to basic quantum theory *Pauling and Wilson* and *Kauzmann* are extremely readable, while the books by *Dirac*, by *Merzbacher*, by *Slater and Frank*, by *Schiff*, and by *Eyring*, *Walter and Kimball* are all to be recommended. There are also relevant parts of the many volume series by *Landau and Lifschitz*. Thermal equilibrium is treated in standard books on thermodynamics and statistical mechanics: treatments with strong sections on the principles and philosophy are those by *Tolman*, by *Hinshelwood*, and by *Fowler and Guggenheim*. Molecular energy relaxation features in general are treated by *Cottrell and McCoubrey* and the nuclear relaxation features by *Slichter* and by *Abragam*.

Electromagnetic radiation is covered by *Heitler* and by *Born and Wolf* as well as in sections of *Landau and Lifschitz* and of *Slater and Frank* and many textbooks on physics. Special aspects of selection rules and group theory are covered in many of the texts mentioned, but specially on these topics are the translation of *Wigner's Gruppentheorie* and the book by *Heine*. Line widths feature prominently in *Breene* and in *Penner*.

Nuclear resonance is treated as regards its general aspects by *Slichter*, by *Carrington and McLachlan* and by *Abragam*, while the early broad line work is reviewed by *Andrew*. The standard high resolution work is by *Pople*, *Bernstein and Schneider* and a treatment with emphasis on organic chemistry is that of *Jackman*. There is a comprehensive treatment by *Emsley, Feeney and Sutcliffe*, and a shorter book by *Lynden-Bell and Harris*. Quadrupole resonance is the particular concern of *Lucken* and of *Das and Hahn* and the interpretation of quadrupole coupling is covered by *Townes and Schawlow*.

The standard background work on magnetism by *van Vleck* was written before successful resonance experiments were performed. *Pake* and *Carrington and McLachlan* provide a useful introduction to electron resonance and parts of *Abragam* are also relevant. The instrumental side and early work on application to radicals, as opposed to transition metals, is given by *Ingram*. For transition and rare earth metals one should refer to *Bleaney and Stevens* and much of relevance to resonance aspects is to be found in *Griffith* and in *Ballhausen*.

The Mössbauer effect is too recently discovered to be exhaustively treated but there is a book by *Wertheim* and an article by *Boyle and Hall* which are useful.

Gas microwave spectroscopy is covered by *Townes and Schawlow* and by *Gordy, Smith and Trambarulo*. Rotation in liquids is treated by *Smyth*, and by *Hill, Vaughan, Price and Davies*, while the classic monograph by *Debye* has matter that is still relevant.

Infra-red and Raman spectra is actually the title of a work by *Herzberg* (*iii*) and its excellence has deterred competitors. Force constants and related aspects are treated by *Wilson, Decius and Cross* and rotational aspects in the companion volume by *Allen and Cross*. Instrumental aspects of the infra-red are covered in miscellaneous places, not least in instrument makers' manuals, and in *Potts*, while the more fundamental experimental information is to be found in *Smith, Jones and Chasmar*. The standard works for characteristic frequencies are *Bellamy* (*i*) and (*ii*), although a long article by *Jones and Sandorfy* covers some features more thoroughly. The specialist work on inorganic materials is by *Lawson* and on polymeric materials by *Zbinden*. Raman spectra are covered by some of these works and *Hibben* deals specifically with this subject.

Condon and Shortley must be on the shelves of most atomic spectroscopists, while at a more elementary level *Herzberg* (*i*) is an excellent introduction. Intermediate are numerous textbooks including those of *Kuhn* and of *White*. Special high resolution instrumental systems are covered by *Tolansky*. The related transition metal ion spectra are covered by *Griffith* and by *Ballhausen*. *Elwell and Gidley* treat the analytical aspects of atomic absorption.

A simple, but good, introduction to electronic spectra is that of *Bowen* (*i*). *Jaffé and Orchin* and *Scott* are more extensive, while *Murrell* concentrates on simple theoretical features. *Pringsheim*, *Bowen* (*ii*) and *Parker* deal with fluorescence. Optically active compounds are the special concern of *Velluz, Legrand and Grossjean* and of *Djerassi. Herzberg* (*ii*) has written the standard work on diatomic spectra and the companion volume, *Herzberg* (*iv*), is on polyatomic electronic spectra.

In matters of nomenclature the practice of *Herzberg* is generally followed, but some gaps are covered by *Mulliken's* recommendations. *Mathieson* has some problems in organic chemistry aspects.

A. ABRAGAM, *The principles of nuclear magnetism*, Oxford, 1961.

H. C. ALLEN and P. C. CROSS, *Molecular vib-rotors*, Wiley, 1963.

E. R. ANDREW, *Nuclear magnetic resonance*, Cambridge, 1955.

C. J. BALLHAUSEN, *Introduction to ligand field theory*, McGraw-Hill, 1962.

L. J. BELLAMY (i), *The infra-red spectra of complex molecules*, Methuen, 1958.

L. J. BELLAMY (ii), *Advances in infrared group analysis*, Methuen, 1968.

B. BLEANEY and K. W. H. STEVENS, 'Paramagnetic resonance', *Reports on progress in physics*, **16**, 108, 1953.

M. BORN and E. WOLF, *Principles of optics; electromagnetic theory of propagation, interference and diffraction of light*, Pergamon, 1959.

E. J. BOWEN (i), *Chemical aspects of light*, Oxford, 1942.

E. J. BOWEN (ii), Ed., *Luminescence in chemistry*, Van Nostrand, 1968.

A. J. F. BOYLE and H. E. HALL, 'The Mössbauer effect', *Reports on progress in physics*, **25**, 441, 1962.

R. G. BREENE, *The shift and shape of spectral lines*, Pergamon, 1961.

A. CARRINGTON and A. D. MCLACHLAN, *Introduction to magnetic resonance*, Harper Row, 1967.

E. U. CONDON and G. H. SHORTLEY, *The theory of atomic spectra*, Cambridge, 1935.

T. L. COTTRELL and J. C. MCCOUBREY, *Molecular energy transfer in gases*, Butterworths, 1961.

T. P. DAS and E. L. HAHN, *Nuclear quadrupole resonance spectroscopy; Solid state physics*, Supplement 1, Academic Press, 1958.

P. DEBYE, *Polar molecules*, Chemical Catalog Inc., 1929.

P. A. M. DIRAC, *The principles of quantum mechanics*, Oxford, 1947.

C. DJERASSI, *Optical rotatory dispersion*, McGraw-Hill, 1960.

W. T. ELWELL and J. A. F. GIDLEY, *Atomic absorption spectrophotometry*, Pergamon, 1961.

J. W. EMSLEY, J. FEENEY and L. H. SUTCLIFFE, *High resolution nuclear magnetic resonance spectroscopy*, Pergamon, 1966.

H. EYRING, J. WALTER and G. E. KIMBALL, *Quantum chemistry*, Wiley, 1944.

R. H. FOWLER and E. A. GUGGENHEIM, *Statistical thermodynamics*, Cambridge, 1939.

W. GORDY, W. V. SMITH and R. F. TRAMBARULO, *Microwave spectroscopy*, Wiley, 1953.

J. S. GRIFFITH, *The theory of transition metal ions*, Cambridge, 1961.

W. HEITLER, *The quantum theory of radiation*, Oxford, 1954.

V. HEINE, *Group theory in quantum mechanics*, Pergamon, 1960.

G. HERZBERG (i) *Atomic spectra and atomic structure*, Prentice-Hall, 1937.

G. HERZBERG (ii) *Spectra of diatomic molecules*, Van Nostrand, 1950.

G. HERZBERG (iii) *Infra-red and Raman spectra*, Van Nostrand, 1945.

G. HERZBERG (iv) *Electronic spectra of polyatomic molecules*, Van Nostrand, 1966.

J. H. HIBBEN, *The Raman effect and its chemical applications*, Reinhold, 1939.

N. E. HILL, W. E. VAUGHAN, A. H. PRICE and M. M. DAVIES, *Dielectric properties and molecular behaviour*, Van Nostrand, 1969.

C. N. HINSHELWOOD, *The structure of physical chemistry*, Oxford, 1951.

D. J. E. INGRAM, *Free radicals as studied by electron spin resonance*, Butterworths, 1958.

L. M. JACKMAN, *Applications of nuclear magnetic resonance spectroscopy in organic chemistry*, Pergamon, 1959.

H. H. JAFFE and M. ORCHIN, *Theory and applications of ultraviolet spectroscopy*, Wiley, 1962.

R. N. JONES and C. SANDORFY, 'The applications of infra-red and Raman spectrometry to the elucidation of molecular structure', Chapter IV of *Chemical applications of spectroscopy*, ed. W. West, being Vol. IX of *Techniques of organic chemistry*, ed. A. Weissberger, Interscience, 1956.

W. KAUZMANN, *Quantum chemistry*, Academic Press, 1957.

H. G. KUHN, *Atomic spectra*, Longmans, 1962.

L. D. LANDAU and E. M. LIFSHITZ, *Course of theoretical physics*, Pergamon, volumes from 1956.

K. E. LAWSON, *Infra-red absorption of inorganic substances*, Reinhold, 1961.

E. A. C. LUCKEN, *Nuclear quadrupole coupling constants*, Academic Press, 1969.

R. M. LYNDEN-BELL and R. K. HARRIS, *Nuclear Magnetic resonance spectroscopy*, Nelson, 1969.

D. H. MATHIESON, Ed., *Interpretation of organic spectra*, Academic Press, 1965.

E. MERZBACHER, *Quantum mechanics*, Wiley, 1961.

R. S. MULLIKEN, 'Report on notation for the spectra of polyatomic molecules', *Journal of Chemical Physics*, **23**, 1997, 1955.

J. N. MURRELL, *The theory of the electronic spectra of organic molecules*, Methuen, 1963.

G. E. PAKE, *Paramagnetic resonance*, Benjamin, 1962.

C. A. PARKER, *Photoluminescence of solutions*, Elsevier, 1968.

L. PAULING and E. B. WILSON, *Introduction to quantum mechanics*, McGraw-Hill, 1935.

S. S. PENNER, *Quantitative molecular spectroscopy*, Pergamon, 1959.

J. A. POPLE, W. G. SCHNEIDER and H. J. BERNSTEIN, *High resolution nuclear magnetic resonance*, McGraw-Hill, 1959.

W. J. POTTS, *Chemical infra-red spectroscopy*, Wiley, 1963.

P. PRINGSHEIM, *Fluorescence and phosphorescence*, Interscience, 1949.

L. I. SCHIFF, *Quantum mechanics*, McGraw-Hill, 1955.

A. I. SCOTT, *Interpretation of the ultraviolet spectra of natural products*, Pergamon, 1964.

J. C. SLATER and N. H. FRANK, *Introduction to theoretical physics*, McGraw-Hill, 1933.

C. P. SLICHTER, *Principles of magnetic resonance*, Harper & Row, 1963.

R. A. SMITH, F. E. JONES and R. P. CHASMAR, *The detection and measurement of infra-red radiation*, Oxford, 1957.

C. P. SMYTH, *Dielectric behaviour and structure*, McGraw-Hill, 1952.

S. TOLANSKY, *High resolution spectroscopy*, Methuen, 1947.

R. C. TOLMAN, *The principles of statistical mechanics*, Oxford, 1938.

C. H. TOWNES and A. L. SCHAWLOW, *Microwave spectroscopy*, McGraw-Hill, 1955.

L. VELLUX, M. LEGRAND and M. GROSJEAN, *Optical circular dichroism*, Academic Press, 1965.

J. H. VAN VLECK, *The theory of electric and magnetic susceptibilities*, Oxford, 1932.

G. K. WERTHEIM, *Mössbauer effect; principles and applications*, Academic Press, 1964.

H. E. WHITE, *Introduction to atomic spectra*, McGraw-Hill, 1934.

E. P. WIGNER, *Group theory and its application to the quantum mechanics of atomic specta*, Academic Press, 1959.

E. B. WILSON, J. C. DECIUS and P. C. CROSS, *Molecular vibrations*, McGraw-Hill, 1955.

R. ZBINDEN, *Infra-red spectroscopy of high polymers*, Academic Press, 1964.

Acknowledgements

The detailed numerical and graphical examples have been adapted from the original literature in various ways. In the following cases part or all of the material used was based on the work cited.

Figure 3.8 J. N. Shoolery, *Disc. Far. Soc.*, **19** (1955).
Figure 3.9 C. E. Looney, W. D. Phillips and E. L. Reilly, *J. Amer. Chem. Soc.*, **79,** 6136 (1957).
Figure 3.10 R. A. Ogg and J. D. Ray, *Disc. Far. Soc.*, **19,** 239 (1955).
Figure 3.1 and Problem 8 G. E. Pake, *J. Chem. Phys.*, **16,** 327 (1948).
Figure 5.6 J. E. Wertz and J. L. Vivo, *J. Chem. Phys.*, **23,** 2441 (1955).
Figure 5.8 N. M. Atherton and D. H. Whiffen, *Mol. Phys.*, **3,** 1 (1960).
Figure 5.9 C. A. Hutchison and B. W. Mangum, *J. Chem. Phys.*, **34,** 908 (1961).
Figure 6.3 W. Kerler and W. Neuwirth, *Mössbauer Effect*, ed. D. M. J. Compton and A. H. Schoen, Wiley (1962), p. 90.
Figure 7.4 C. H. Townes, A. N. Holden, J. Bardeen and F. R. Merritt, *Phys. Rev.*, **71,** 644 (1947).
Figure 7.6 D. H. Whiffen, *Trans. Far. Soc.*, **46,** 130 (1950).
Figure 8.2 S. A. Barker, E. J. Bourne, R. M. Pinkard and D. H. Whiffen, *J. Chem. Soc.*, 807 (1959).
Figure 8.5 Based on a problem presented by J. C. Tatlow, *et al.*, Chemistry Department, Birmingham University.
Figure 9.2 J. H. S. Green, W. Kynaston and H. M. Paisley, *Spectrochim. Acta*, **19,** 549 (1963).
Figure 10.3 E. K. Plyler and E. D. Tidwell, *Z. Electrochem.*, **64,** 717 (1960).
Figure 10.5, Tables 10.1 and 10.2 H. J. Bernstein and D A. Ramsay, *J. Chem. Phys.*, **17,** 556 (1949); H. J. Bernstein and A. D. E. Pullin, *Can. J. Chem.*, **30,** 963 (1952); K. S. Pitzer and J. J. Hollenberg, *J. Amer. Chem. Soc.*, **76,** 1493 (1954).

Figure 11.1, Tables 11.1 and 11.2 and Problem 27 C. E. Moore, Atomic Energy Tables, *Nat. Bur. Stand. Circ.* **467** (1949–58).

Figure 12.4 G. Kortum and B. Finckh, *Z. Phys. Chem.*, **52B,** 263 (1942).

Table 6.1 O. C. Kistner, V. Jaccarino and L. R. Walker, *Mössbauer Effect*, ed. D. M. J. Compton and A. H. Schoen, Wiley (1962), p. 264.

Tables 7.1 and 7.2 C. H. Townes and A. L. Schawlow, *Microwave Spectra*, McGraw-Hill (1955), Appendix VI and references quoted therein.

Table 12.2 K. W. Hausser, R. Kuhn, A. Smakula and M. Hoffer, *Z. Phys. Chem.*, **29B,** 371 (1935).

Table 12.3 A. E. Gillam and E. S. Stern, *Electronic Absorption Spectroscopy*, Arnold (1954), Table 9.3.

Table 13.1 B. Eisler and R. F. Barrow, *Proc. Phys. Soc.*, **62A,** 740 (1949).

Problem 7 J. N. Shoolery, quoted by J. A. Pople, W. G. Schneider and H. J. Bernstein, *High Resolution Nuclear Magnetic Resonance*, McGraw-Hill (1959), p. 335.

Problem 9 D. W. Moore, *J. Chem. Phys.*, **34,** 1470 (1961).

Problem 10 H. G. Dehmelt, *Z. Phys.*, **130,** 356 (1951); K. Shimomura, *J. Sci. Hiroshima Univ.*, **17A,** 383 (1954).

Problem 11 R. Livingston, *J. Chem. Phys.*, **20,** 1170 (1952).

Problem 12 R. W. Fessenden and R. H. Schuler, *J. Chem. Phys.*, **33,** 935 (1960).

Problem 13 D. Pooley and D. H. Whiffen, *Trans. Far. Soc.*, **57,** 1445 (1961).

Problem 14 R. H. Herber and G. K. Wertheim, *Mössbauer Effect*, ed. D. M. J. Compton and A. H. Schoen, Wiley (1962), p. 105.

Problem 18 F. J. Levin and J. G. Winans, *Phys. Rev.*, **84,** 431 (1951).

Problem 19 D. H. Whiffen and H. W. Thompson, *Trans. Far. Soc.*, **42A,** 122 (1946).

Problem 20 R. N. Jones and C. Sandorfy, *Techniques of Organic Chemistry*, ed. A. Weissberger, Vol. IX, *Chemical Applications of Spectroscopy*, ed. W. West, Chapter IV, Figures 58, 59 and 60.

Problem 22 H. D. Babcock and L. Herzberg, *Astrophys. J.*, **108,** 167 (1948).

Problem 25 B. P. Stoicheff, *Can. J. Phys.*, **32,** 630 (1954).

Problem 29 R. F. Barrow and M. F. R. Mulcahy, *Proc. Phys. Soc.*, **61,** 99 (1948).

*Gs

Symbols used

A Coefficient of spontaneous emission
Rotational constant
Symmetry class
as Antisymmetric
a Parameter in Morse function
Symmetry class
Linear dimension
Hyperfine coupling constant
B Coefficient of absorption and induced emission
Magnetic induction
Rotational constant; especially linear molecules
Symmetry class
b Symmetry class
Linear dimension
C Rotational constant
Angular momentum function defined on p. 181
C_{2v} A point group to which cis-dichlorethylene belongs
c Velocity of electromagnetic waves in vacuo
Spin-rotation interaction constant
Linear dimension
Concentration
D Designation for atomic state with $L=2$
D Zero-field splitting parameter
Dissociation energy
Centrifugal distortion constant
D_{2h} A point group to which ethylene belongs
d Differential operator
d electron or d orbital, case where $l=2$
dp Depolarized

d Grating spacing
E Electric field
Energy
Zero-field splitting parameter
e As subscript, equilibrium value
e Charge on the electron
F Atomic state designation for $L=3$
F Total angular momentum[1] including nucleus
f f electron or f orbital, case where $l=3$
g As subscript, ground state
g Degeneracy
Electron magnetogyric ratio in Bohr magnetons
Especially as subscript, gerade
H Magnetic field
h Planck constant. $\hbar=h/2\pi$
I Moment of inertia
Radiation density per unit area
Nuclear spin[1]
i Square root of minus one
i Identifying subscript
J Nuclear spin-spin coupling constant
Angular momentum,[1] usually total excluding nuclear spin
j One electron total angular momentum[1]
Identifying subscript
K Extinction coefficient
Angular momentum[1] about symmetry axis
k Gas constant per molecule
Force constant
Rate constant
L Total electron orbital angular momentum[1]
l One electron orbital angular momentum[1]
Cell length
M Component of angular momentum along field or z axis
m Medium intensity
m Mass
m_e Electron mass
m_p Proton mass
N_A Avogadro constant
N A general integer
Number per unit volume
n General non-bonding orbital
n Order of grating

n Principal atomic quantum number
Refractive index
State population
o As subscript, appropriate ground state average
P Designation for atomic state with $L=1$
P Dielectric polarization
p p electron or p orbital, case when $l=1$
Polarized
pp Partially polarized
p Electric dipole moment
Q Ratio of hyperfine coupling to spin population
Nuclear electric quadrupole moment
Partition function
q Minus the electric field gradient
General displacement coordinate
R_x Rotation about axis written as subscript
R Nuclear radius
Gas constant per mole
Rydberg constant
r Distance variable, radial in atoms, internuclear in molecules
S Designation for atomic state with $L=0$
S Total electron spin angular momentum[1]
s s electron or s orbital, case where $l=0$
Symmetric
s One electron spin angular momentum[1]
T_x Translation in direction indicated by subscript
T Absolute temperature
Nuclear relaxation time
t Time variable
U_{p} Unbalanced p electrons; see p. 50
u Especially as subscript, ungerade
Velocity
V Electric potential
Potential energy
v Very, as in vs very strong
v Vibrational quantum number
w Weak intensity
X Ground electronic state
X Zero field splitting parameter
x Cartesian coordinate
Anharmonicity coefficient
Y Zero-field splitting parameter

y Cartesian coordinate
Cubic anharmonicity coefficient
Z Zero-field splitting parameter
Atomic number or nuclear charge
z Cartesian coordinate
α Fine structure constant
Polarizability
Vibration-rotation interaction constant
Attenuation coefficient
γ Magneto-gyric ratio of nuclei
See also γ^2 as related to the anisotropy of the polarizability, p. 119
Δ Change of, especially in selection rules
Designation for state of linear molecule with $\Lambda = 2$
δ Small increment
Chemical shift
∂ Partial derivative
ε Molecular extinction coefficient
Dielectric permittivity
ε_0 Permittivity of a vacuum
ζ One electron spin own-orbit coupling parameter
η Asymmetry parameter in quadrupole coupling
θ Angular variable; especially polar angle
κ Spin-spin coupling in solid
Λ Axial angular momentum[1] in linear molecules
λ Radiation wavelength
μ Magnetic moment[1] especially of nuclei
μ_B Bohr magneton
μ_N Nuclear magneton
μ_0 Permeability of a vacuum
ν Frequency
Π Product sign
Designation for state of linear molecule with $\Lambda = 1$
π π electron or π orbital, case where there is a nodal plane coinciding with a molecular plane or containing the axis of a linear molecule
ρ Radiation density
Nuclear charge density
Depolarization ratio
Spin population
Σ Summation sign
Designation for state of linear molecule with $\Lambda = 0$
Projection of electronic spin on axis of linear molecule
σ Wave number

τ Relaxation time
Chemical shift on tau scale
$d\tau$ Volume element in an integral
φ Angular variable, especially azimuthal angle
One electron wave function
ψ Wave function generally
Ω Axial projection of orbital plus spin angular momentum[1] in a linear molecule
ω Frequency in radians per second. $\omega = 2\pi\nu$

[1] For angular momenta, etc.; the same symbol is used for the expectation value or quantum number and for the vector operator.

Units, physical constants, etc.

Definitions (abbreviated)

Metre, m: Length equal to 1 650 763·73 wavelengths in vacuum of krypton-86 radiation of a specified transition.

Kilogram, kg: Mass of international prototype kilogram.

Second, s: Duration of 9 192 631 770 periods of the ground state hyperfine transition of caesium-133.

Ampere, A: That electric current which maintained in two thin, straight, infinite conductors one metre apart in a vacuum produces a force of 2×10^{-7} m kg s^{-2} per metre of length.

Kelvin, K: Unit of temperature of $(273{\cdot}16)^{-1}$ times the thermodynamic temperature of the triple point of water.

Mole, mol: Amount of substance containing as many specified elementary units as there are carbon atoms in 0·012 kg of carbon-12.

CO DATA (1970) recommended values of the physical constants and some other useful quantities Those given exact values are a matter of definition, whereas the remainder may be modified by experiments performed since 1969 and new recommended values are likely to be promulgated in the future; this list is based on B. N. Taylor, W. H. Parker and D. N. Langenburg, *Rev. Mod. Phys.* **41,** 375 (1969) which reference should be unsuited for limits of error.

μ_0	Permeability of a vacuum	*exactly* $4\pi \times 10^{-7}$ m kg s^{-2} A^{-2}
ε_0	Permittivity of a vacuum ($=\mu_0^{-1} c^{-2}$):	$8{\cdot}854\,185\,3 \times 10^{-12}$ m^{-3} kg^{-1} s^4 A^2
c	Speed of light in a vacuum:	$2{\cdot}997\,925\,0 \times 10^{8}$ m s^{-1}
e	Electric charge of a proton:	$1{\cdot}602\,191\,7 \times 10^{-19}$ s A
h	Planck constant:	$6{\cdot}626\,196 \times 10^{-34}$ m^2 kg s^{-1}
$\hbar$	$=h/2\pi$:	$1{\cdot}054\,591\,9 \times 10^{-34}$ m^2 kg s^{-1}

k	Boltzmann constant:	$1{\cdot}380\,622 \times 10^{-23}$ m^2 kg s^{-2} K^{-1}
N_A	Avogadro constant:	$6{\cdot}022\,169 \times 10^{23}$ mol^{-1}
R	Gas constant ($=N_A\,k$):	$8{\cdot}314\,34$ m^2 kg s^{-2} K^{-1} mol^{-1}
F	Faraday constant:	$9{\cdot}648\,670 \times 10^4$ s A mol^{-1}
m_e	Mass of the electron:	$9{\cdot}109\,558 \times 10^{-31}$ kg
m_p	Mass of the proton:	$1{\cdot}672\,614 \times 10^{-27}$ kg
G	Gravitational constant:	$6{\cdot}673\,2 \times 10^{-11}$ m^3 kg^{-1} s^{-2}
α	Fine structure constant ($=\mu_0 e^2 c/2h$):	$7{\cdot}297\,351 \times 10^{-3}$
a_0	Bohr radius ($=h^2/\pi\,\mu_0\,c^2\,m_e\,e^2$):	$5{\cdot}291\,771\,5 \times 10^{-11}$ m
μ_B	Bohr magneton ($=e\,\hbar/2\,m_e$):	$9{\cdot}274\,096 \times 10^{-24}$ m^2 A
μ_N	Nuclear magneton ($=e\,\hbar/2\,m_p$):	$5{\cdot}050\,951 \times 10^{-27}$ m^2 A
R_∞	Infinite Rydberg ($=\mu_0^2\,m_e\,e^4\,c^3/8\,h^3$):	$1{\cdot}097\,373\,12 \times 10^7$ m^{-1}
	Rydberg energy ($=h\,c\,R_\infty$):	$2{\cdot}179\,914 \times 10^{-18}$ m^2 kg s^{-2}
	Standard atmosphere	*exactly* $1{\cdot}013\,25 \times 10^5$ m^{-1} kg s^{-2}
g	Standard acceleration of gravity	*exactly* $9{\cdot}806\,65$ m s^{-2}

SI units with special names

frequency	hertz	Hz	s^{-1}
energy	joule	J	m^2 kg s^{-2}
force	newton	N	m kg s^{-2}
power	watt	W	m^2 kg s^{-3}
pressure	pascal	Pa	m^{-1} kg s^{-2}
electric charge	coulomb	C	s A
electric potential difference	volt	V	m^2 kg s^{-3} A^{-1}
electric resistance	ohm	Ω	m^2 kg s^{-3} A^{-2}
electric capacitance	farad	F	m^{-2} kg^{-1} s^4 A^2
magnetic induction or flux density	tesla	T	kg s^{-2} A^{-1}
magnetic flux	weber	Wb	m^2 kg s^{-2} A^{-1}
magnetic inductance	henry	H	m^2 kg s^{-2} A^{-2}

Other units

Common temperature t in degrees celsius defined by $t/°\mathrm{C}=(T/\mathrm{K})-273{\cdot}15$
electron volt eV $=1{\cdot}602\,191\,7 \times 10^{-19}$ m^2 kg s^{-2} or J
this is equivalent to $9{\cdot}648\,670 \times 10^4$ m^2 kg s^{-2} mol^{-1} or J mol^{-1}
unified atomic mass unit u $=1{\cdot}660\,531 \times 10^{-27}$ kg
thermochemical calorie cal $=$ *exactly* $4{\cdot}184$ kg m^2 s^{-2} or J

Some useful, very approximate, equivalences

1 cm$^{-1} \equiv 3 \times 10^{10}$ Hz $\equiv 12$ J mol^{-1}
1 eV $\equiv 8000$ cm$^{-1} \equiv 100$ kJ mol^{-1}

Index

D

E

F

G

H

S

T

U

V